The Urban Book Series

The Urban Book Series is a resource for urban studies and geography research worldwide. It provides a unique and innovative resource for the latest developments in the field, nurturing a comprehensive and encompassing publication venue for urban studies, urban geography, planning and regional development.

The series publishes peer-reviewed volumes related to urbanization, sustainability, urban environments, sustainable urbanism, governance, globalization, urban and sustainable development, spatial and area studies, urban management, transport systems, urban infrastructure, urban dynamics, green cities and urban landscapes. It also invites research which documents urbanization processes and urban dynamics on a national, regional and local level, welcoming case studies, as well as comparative and applied research.

The series will appeal to urbanists, geographers, planners, engineers, architects, policy makers, and to all of those interested in a wide-ranging overview of contemporary urban studies and innovations in the field. It accepts monographs, edited volumes and textbooks.

Indexed by Scopus.

Isabel Ruiz-Mallén • Hug March
Mar Satorras
Editors

Urban Resilience to the Climate Emergency

Unravelling the transformative potential of institutional and grassroots initiatives

Springer

Editors
Isabel Ruiz-Mallén
Estudis de Psicologia i Ciències de l'Educació and Internet Interdisciplinary Institute (IN3)
Universitat Oberta de Catalunya (UOC)
Barcelona, Spain

Hug March
Estudis d'Economia i Empresa and Internet Interdisciplinary Institute (IN3)
Universitat Oberta de Catalunya (UOC)
Barcelona, Spain

Mar Satorras
Internet Interdisciplinary Institute (IN3)
Universitat Oberta de Catalunya (UOC)
Barcelona, Spain

Barcelona Institute of Regional and Metropolitan Studies (IERMB)
Universitat Autònoma de Barcelona
Bellaterra, Spain

ISSN 2365-757X ISSN 2365-7588 (electronic)
The Urban Book Series
ISBN 978-3-031-07300-7 ISBN 978-3-031-07301-4 (eBook)
https://doi.org/10.1007/978-3-031-07301-4

This Springer imprint is published by the registered company Springer Nature Switzerland AG
The registered company address is: Gewerbestrasse 11, 6330 Cham, Switzerland

Foreword

When we realized that sustainability was everywhere but too holistic to be effective in its implementation, we chose resilience to be more practical. However, we ended up lost in a mess of interpretations and fuzziness.

This is what I would say as an academic, in 2022, after more than 15 years of in-depth work on urban resilience. However, when introduced back in 2009 (when Peter Newman published his book *Resilient cities: Responding to peak oil and climate change*), or even earlier in 2005 (when Vale and Campanella published the book *The resilient city: How modern cities recover from disaster*), resilience was a clear answer to socio-environmental crises, driving toward a safe and green urban future. When strategically supporting Rockefeller Foundation in the running for its goal of launching the 100 Resilient Cities program, Judith Rodin reported resilience as something bringing clear principles of adaptation, learning, and robustness, and thus supporting cities in a genuinely pragmatic way, toward action. And indeed, even we, from the academic lens, perceived that the message of resilience is more radical for policy makers than that of sustainability, as reported by Brand and Jax in 2007 while exploring the meanings of urban resilience as a boundary object.

Nevertheless, it took no more than a few years to realize how far resilient agendas and discourses were from something clear and desirable. The very same Lawrence J. Vale, in 2014, was indeed publishing *The politics of resilient cities: Whose resilience and whose city*, after a bunch of us from different disciplines were publishing in 2012 critical essays questioning whether resilience was something driving toward change (sustainable change and transformations) or keeping the status quo by fostering robustness of our built environment and infrastructures, thus keeping our socio-economic and political systems under the same behaviors, locking us to increasing risks.

If one looks back, these 10 years have been full of quite exciting perspectives around resilience and transformations, resilience and sustainability, and resilience and justice, all trying to situate resilience while recognizing its extreme fuzziness, nested within its metaphorical meaning of adaptive capacity, and lack of consistent implementation guidelines with respect to justice and sustainability.

This book is a wonderful brick of thoughts and examples under the light of a political ecology eye, fitting this path from the (blind, or unfortunately mainstream) normative-positive understanding of resilience toward a more critical perspective about how resilience should be carefully framed (and how it should not). Grounding on the role and meanings of community and institutions, and looking at resilience from top-down and bottom-up lenses, it addresses and develops interesting reflections on the gaps existing between urban transformations and the understanding of what is community, and how to overcome uncritical assumptions around the universal benefits of resilience. The need for a greater understanding of adaptive and integrated governance processes is reflected throughout the chapters, as the editors nicely introduce within the book's preface. Case studies allow the readers to explore and understand power, benefits, capacities, (mal)adaptations, and the complexity of dealing with urban transformations through the lens of political ecology. The evidence presented in the book help to structure a deeper shared understanding of how to filter, unveil, and organize, and the many and different potential facets of resilience when thinking of its implementation.

When ecologists were exploring resilience and jumped from Buzz Holling's first essay on plants and bugs to the resilience of social-ecological systems, the Resilience Alliance was established as a global network of researchers sharing knowledge and gathering case studies on tipping points and thresholds. The goal was to build the foundations for a common understanding of coping and regulatory mechanisms for ecosystems to avoid collapses and informing our governance for keeping social-ecological systems within sustainable equilibrium states. This empirical body of knowledge, around social-ecological systems' dynamics, confers now to (social-ecological) resilience thinking a scientific flavor, which urban resilience is still missing because highly politicized. This book is raising a greater and more critical understanding of urban resilience, toward the much needed multidisciplinary and critical lenses, exploring many hidden facets of this concept.

Barcelona, Spain Lorenzo Chelleri

Preface

In the context of increasing importance of the *local* in addressing global socio-environmental challenges, such as climate change, it is imperative to go beyond mainstream incremental visions of urban resilience and shift the attention to more transformative ways of building resilience, including those initiatives from the grassroots and those based on co-production processes with city government authorities.

With this motivation, as editors, we looked for academic colleagues who were exploring different forms of urban climate governance in cities while putting the focus on their potential for transformation through specific case studies. Our aim was to advance the state of the art in urban climate resilience covered by previous publications by providing a more nuanced understanding of the transformational opportunities and challenges underlying the development of urban resilience strategies dealing with climatic changes guided by the lens of political ecology, environmental justice, and participatory governance.

Given the multiple – and even conflicting – meanings of urban resilience both in academic and political debates on resilience and specifically in the literature on urban resilience, this book does not stick to one specific approach to urban climate resilience, but rather on how resilience is understood, planned, and performed by those developing and implementing practical actions on the ground with a focus on transformation.

The book starts with two chapters, in Part I, addressing the main international debates on urban resilience to the climate emergency from the lenses of political ecology, environmental justice, and participatory governance, with a special emphasis on urban climate resilience and transformation. Then the book dives into several case studies through two different sections. In Part II, three chapters analyse evidence from the Global North and South (i.e., Europe and Latin America, respectively) to make progress in the understanding of the (un)expected implications of top-down resilience plans to deal with the impacts of climate change in cities, especially on the most vulnerable and marginalised urban populations. In Part III, four chapters relying on data from different Central and Southern European cities (i.e.,

Paris, Madrid, Barcelona, Seville) show the critical role of grassroots organisations and individual citizens in promoting alternative and transformative strategies of urban climate resilience by means of co-produced and self-organised initiatives. Finally, at the end of the book, the afterword chapter disentangles the different ideas, perspectives, and approaches of transformation showcased by the compiled case studies, as well as their opportunities and challenges, to reflect on the ways forward in order to build urban resilience from and within transformative learning processes. All chapters are original work except Chap. 4, which is reused from an article published in Elsevier in 2020.[1]

As editors, we carefully selected the contributions included in this book by favouring diverse, original, and groundbreaking perspectives to address this topic, always stressing a qualitative approach, and limiting the number of case studies to a manageable amount in order to put depth and quality before quantity. Because of the geographical bias of the three of us (who are working on climate resilience in the Barcelona region), some of the chapters draw from Barcelona, which should not be understood as a limitation, but as an opportunity to explore the transformational aspects of urban climate resilience in a city that has promised to support both institutional and grassroots efforts in this field. All in all, we selected paradigmatic case studies that provide evidence and insights relevant for international academic and policymaking debates.

Therefore, this volume compiles critical research documenting the articulation of urban resilience strategies that deal with climatic changes and hydro-climatic risks in different urban areas, mostly European ones, which emphasise transformative processes, capacities, and outcomes dealing with power issues and/or highlighting social inequalities in urban planning for climate change. It has done so by accompanying the empirical evidence with conceptual and theoretical approaches to contribute to the critical understanding of urban resilience strategies to the climate emergency. The critical view on the topic of urban resilience and its interdisciplinary nature makes the book appealing for academic and educational purposes, as well as for supporting decision-making in urban planning. The three editors hope that its contents inspire existing debates and reflection for framing new ideas and research hypothesis, while illuminating research, teaching, practice, and policymaking in urban climate resilience.

Finally, we are grateful for financial support from the Spanish Ministry of Science and Innovation through the RESCITIES project (*La ecología política de la resiliencia urbana a los fenómenos hidro-climáticos en España*): Plan Nacional (PGC2018-100996-A-I00 (MCIU/AEI/FEDER, UE)) that made this book possible. We also thank the Spanish State Research Agency through a "Juan de la Cierva-Formación" programme (FJCI-2017-31723) as well as the grant RYC-2015-17676

[1] Reprinted from Cities, 99, Zografos C, Klause KA, Connolly JJT, Anguelovski I, The everyday politics of urban transformational adaptation: Struggles for authority and the Barcelona superblock project, 102613, Copyright (2020), with permission from Elsevier

funded by MCIN/AEI/ https://doi.org/10.13039/501100011033 and by "ESF Investing in your future". We would like to acknowledge the useful comments and suggestions of three anonymous external reviewers in an early stage of the book proposal, which helped us to fully shape the present book. We would also like to thank Raquel Colacios who assisted in the editing process.

Estudis de Psicologia i Ciències de l'Educació
and Internet Interdisciplinary Institute (IN3)
Universitat Oberta de Catalunya (UOC), Barcelona, Spain

Isabel Ruiz-Mallén

Contents

Contributors

Editors

Isabel Ruiz-Mallén Estudis de Psicologia i Ciències de l'Educació and Internet Interdisciplinary Institute (IN3), Universitat Oberta de Catalunya (UOC), Barcelona, Spain

Hug March Estudis d'Economia i Empresa and Internet Interdisciplinary Institute (IN3), Universitat Oberta de Catalunya (UOC), Barcelona, Spain

Mar Satorras Internet Interdisciplinary Institute (IN3), Universitat Oberta de Catalunya (UOC) and Barcelona Institute of Regional and Metropolitan Studies (IERMB), Universitat Autònoma de Barcelona (UAB), Barcelona, Spain

Authors

Isabelle Anguelovski ICREA, Barcelona Lab for Urban Environmental Justice and Sustainability, Institute of Environmental Science and Technology (ICTA), Universitat Autònoma de Barcelona, Cerdanyola del Vallès, Spain

Pere Ariza-Montobbio Latin American Faculty of Social Sciences (FLACSO), Quito, Ecuador

Francesc Baró Department of Geography and Department of Sociology, Vrije Universiteit Brussel (VUB), Brussels, Belgium

Institute of Environmental Science and Technology (ICTA), Universitat Autònoma de Barcelona, Cerdanyola del Vallès, Spain

Luis Berraquero Díaz Social Research and Participatory Action Group (GISAP), Universidad Pablo de Olavide, Seville, Spain

David A. Camacho Institute of Environmental Science and Technology (ICTA), Universitat Autònoma de Barcelona, Cerdanyola del Vallès, Spain

Andrea Carrión Latin American Faculty of Social Sciences (FLACSO), Quito, Ecuador

Instituto de Altos Estudios Nacionales (IAEN), Quito, Ecuador

James J. T. Connolly School of Community and Regional Planning, University of British Columbia, Vancouver, Canada

Barcelona Lab for Urban Environmental Justice and Sustainability, Barcelona, Spain

Gian Carlo Delgado-Ramos Center for Interdisciplinary Research in the Sciences and Humanities, National Autonomous University of Mexico (UNAM), Mexico City, Mexico

Pablo García-Serrano Serendipia Educativa, Madrid, Spain

Jordi Honey-Rosés Institute for Environmental Science and Technology (ICTA), Universitat Autònoma de Barcelona, Cerdanyola del Vallès, Spain

Kai A. Klause Barcelona Lab for Urban Environmental Justice and Sustainability, Institute of Environmental Science and Technology (ICTA), Universitat Autònoma de Barcelona, Cerdanyola del Vallès, Spain

Ángela Lara García Department of Human Geography, University of Seville, Seville, Spain

Leandro del Moral Ituarte Department of Human Geography, University of Seville, Seville, Spain

Iago Otero Interdisciplinary Centre for Mountain Research, University of Lausanne, Lausanne, Switzerland

Carmen Pérez del Pulgar Environmental Policy Department, Helmholtz Centre for Environmental Research – UFZ, Leipzig, Germany

Department of Political Science, Friedrich-Schiller-University Jena, Jena, Germany

Institute of Environmental Science and Technology (ICTA), Universitat Autònoma de Barcelona, Cerdanyola del Vallès, Spain

Kaitlin Strange Internet Interdisciplinary Institute (IN3), Universitat Oberta de Catalunya (UOC), Barcelona, Spain

Erik Swyngedouw School of Environment, Education and Development, The University of Manchester, Manchester, UK

Christos Zografos Department of Political and Social Sciences, Universitat Pompeu Fabra, Barcelona, Spain

Editors Biography

Isabel Ruiz-Mallén is an associate professor at the Faculty of Psychology and Education Sciences and a senior researcher at the Internet Interdisciplinary Institute (IN3) - Universitat Oberta de Catalunya (UOC), Spain. She is currently leading a research line on grassroots co-creation for sustainable solutions at the Laboratory of Urban Transformation and Global Change (TURBA). She has a proven ability to work across disciplines, from sustainability education to community-based conservation and climate change adaptation and resilience, in multicultural contexts (global North/South, rural/urban), and through exploring innovative participatory approaches (arts-based). Isabel is an environmental scientist holding a MSc in Biological Sciences from UNAM (2005) and a PhD in Environmental Sciences from UAB (2009).

Hug March is an associate professor in the Faculty of Economy and Business, Universitat Oberta de Catalunya (UOC). He is a researcher at the Laboratory of Urban Transformation and Global Change (TURBA, UOC), where he leads a research line on the political ecology of socio-environmental transformation. Hug obtained a PhD in environmental studies from Universitat Autònoma de Barcelona in 2010. He has authored over 60 publications in peer-reviewed journals and book chapters in the fields of environmental studies, urban studies, geography, and water studies. He has done extensive research on the urban political ecology of the water, including research on financialisation and remunicipalisation.

Mar Satorras is a researcher at the Barcelona Institute of Regional and Metropolitan Studies (IERMB), Universitat Autònoma de Barcelona (UAB), and a member of the Laboratory of Urban Transformation and Global Change (TURBA) at IN3-Universitat Oberta de Catalunya (UOC). She is an anthropologist and environmental scientist holding a MSc and PhD in Environmental Studies from UAB. Mar's research focuses on urban sustainability challenges approached under socio-political lenses, with particular interests on the water cycle and the climate emergency. She has conducted research on urban climate adaptation and resilience, by exploring the ways how past and present societies experience and respond to hydro-climatic dynamics. Her interdisciplinary research at the interface between environmental and social sciences combines ethnographic, historical, and geographical perspectives.

Authors Biography

Isabelle Anguelovski is the director of the Barcelona Lab for Urban Environmental Justice and Sustainability (BCNUEJ) and an ICREA Research Professor in the Institute of Environmental Science and Technology at Universitat Autònoma de

Barcelona (ICTA-UAB). She obtained a PhD in urban studies and planning from MIT before returning to Europe in 2011 with a Marie Curie International Incoming Fellowship. Situated at the intersection of urban planning and policy, social inequality, and development studies, her research examines the extent to which urban plans and policy decisions contribute to more just, resilient, healthy, and sustainable cities, and how community groups in distressed neighbourhoods contest the existence, creation, or exacerbation of environmental inequities as a result of urban (re)development processes and policies.

Pere Ariza-Montobbio is an agro-ecologist and environmental scientist whose action research revolves around alternative land use planning, energy metabolism, socio-ecological resilience and transitions, as well as sustainable land management (SLM), with special emphasis on agroecology and regenerative agriculture. During 2020 and 2021, he coordinated the applied research area of the project "Building leadership for the cities of Latin America and the Caribbean in the face of climate change", implemented by FLACSO Ecuador, with funding from IDRC, Canada. His latest research has been on urban political ecology and climate change in Latin America. Currently, Pere combines his research activities with the management of an agroecological farm in Manabí, Ecuador. He participates in social movements and projects promoting agroecology and ecological restoration, as well as popular environmental education.

Francesc Baró is an urban environmental scientist currently appointed as Assistant Professor of Urban Ecology at Vrije Universiteit Brussel (VUB), Belgium, and he is also an affiliated researcher at the Institute of Environmental Science and Technology (ICTA-UAB), Spain. His research is motivated by making cities inclusive, healthy, liveable, resilient, and sustainable for all. In order to understand the complexity of urban social-ecological systems, he conducts applied research at the interface of urban forestry, urban geography, urban planning, and ecosystem service science. His interdisciplinary research approach combines geospatial modelling, advanced quantitative and qualitative data analysis, and participatory methods.

Luis Berraquero Díaz is a researcher and activist. He holds an MSc in applied environmental social research as well as an MSc in sustainability science for managing global change, and he is a doctoral student in environment and society at Pablo de Olavide University. His research focuses on social movement analysis, urban sustainability, urban resilience, and neoliberalisation processes in cities, such as gentrification and touristification. To address these issues, he uses qualitative methods such as ethnography, participatory action research, and power relations analysis. Luis is executive committee member of the Urban Resilience Research Network, member of the Iberian Environmental Anthropology Network, and member of the Ibero-American network of research on urban policies, conflicts, and movements.

David A. Camacho is an urban ecologist working as a pre-doctoral researcher at the Institute of Environmental Science and Technology (ICTA-UAB), Spain. His research focuses on the holistic evaluation of nature-based solutions considering social vulnerabilities, environmental burdens, and ecosystem services by employing both qualitative and quantitative methods.

Andrea Carrión is a geographer specialising in political economy, holding a tenure-track position as an associate professor at the Instituto de Altos Estudios Nacionales del Ecuador (IAEN). Between 2019 and 2021, she coordinated the project "Building leadership for Latin American and Caribbean cities in a changing climate", implemented by FLACSO Ecuador, with funding from IDRC, Canada. Her professional and academic work revolves around planning, land management, public policies, climate resilience, and the social production of habitat. She gives emphasis to documenting praxis, highlighting the differential involvement of multiple actors and everyday processes shaping local governance.

James J. T. Connolly is codirector and Affiliated Researcher of BCNUEJ and Assistant Professor of Community and Regional Planning at the University of British Columbia. Previously he was Assistant Professor of Public Policy and Political Science at Northeastern University and obtained a PhD in urban planning from Columbia University where his research was supported by the United States Department of Housing and Urban Development. His research examines how cities are made greener and more socially just at the same time. His publications examine green gentrification and land use politics, particularly the dynamics of coalition building across community development and mainstream environmental coalitions. He is interested in the spatial and political structure of institutions that shape urban environmental land use policy and how these are changed.

Gian Carlo Delgado Ramos is a researcher at the Center for Interdisciplinary Research on the Sciences and Humanities of the National Autonomous University of Mexico. His main lines of research are urban adaptation and mitigation, urban sustainability and resilience, urban metabolism, urban governance, and urban political ecology. Gian is member of the National Research System of CONACYT and a regular member of the National Academy of Science. He was a lead author in the IPCC's AR5 and a review editor in AR6. He co-chaired UNEP's regional report "The Weight of Cities in Latin America and the Caribbean: Future resource requirements and potential courses of action" (UNEP, 2021).

Pablo García-Serrano has a PhD in urban ecology. His research focuses on participatory design and evaluation for landscape and public space. For the last decade, he has been serving as strategic adviser to develop schoolyard learning environments for public administrations including Spain, Colombia, Costa Rica, and Ecuador.

Jordi Honey-Rosés is a senior researcher at the Institute of Environmental Science and Technology at Universitat Autònoma de Barcelona (ICTA-UAB) and the lead researcher of City Lab Barcelona. He is an environmental planner specialised in urban impact evaluation. He has published widely on urban experiments and impact evaluation in leading international scientific journals and his teaching has been recognised with the prestigious University Killam Teaching Award from the University of British Columbia (UBC). He holds the appointment of honorary research associate at the Institute for Resources, Environment and Sustainability (IRES) at UBC. He has university degrees from the University of California at Berkeley, a Master of Public Policy from the Harvard Kennedy School, and a PhD in regional planning from the University of Illinois, Urbana Champaign.

Kai A. Klause is urban development desk officer in the policy department at the German Catholic Organisation for Development Cooperation, MISEREOR. As such, he advocates for socially and environmentally just urban development in national and global policy discourses. He hereby ties in with his research activities on transformative urban climate governance and the impact of international agricultural trade on urban food security. He holds a postgraduate degree in international cooperation for sustainable development from the Centre for Rural Development/ Humboldt University Berlin, an MSc in environmental, economic, and social sustainability from the Universitat Autònoma de Barcelona (ICTA-UAB), and a BA in development studies from the University of Vienna.

Ángela Lara García is an architect and holds a PhD in geography (University of Seville, Spain), MSc in spatial planning (USA), and MSc in architecture and sustainable urbanism (UNIA, Spain). She is lecturer in the Department of Human Geography (USA). Her research focuses on the social and ecological components of the urban space, especially in relation to the water cycle. She participates in national and international research projects and is member of the Structures and Territorial Systems Research Group (HUM 396), of the Water Poverty Research Network (Waponet), as well as of social organisations such as the New Water Culture Foundation.

Leandro del Moral Ituarte is a full professor in the Department of Geography of Universidad de Sevilla. He has a long and very prestigious academic trajectory on water governance and hydro-climatic risks in Spain. He has participated in several nationally and European Union–funded projects on water management. Between 2013 and 2015, he directed the AQUA-RIBA project that systematised the institutional and technological alternatives for the improvement of the urban water system in Andalusia. He is part of the Board of Trustees of the New Culture of Water Foundation (FNCA), of which he was president (2007–2009). He was a member of the Board of Directors of the Metropolitan Water Company of Seville (EMASESA).

Iago Otero is an environmental scientist working as a researcher at the University of Lausanne, Switzerland. He conducts research on the political ecology of forest fires in relation to transformative change, using qualitative research and participatory methods. He is also interested in the study of inter- and transdisciplinarity and in the relationships between economic growth and biodiversity loss.

Carmen Pérez del Pulgar is a post-doctoral researcher at the Helmholtz-Centre for Environmental Research-UFZ. Her research focuses on how urban socio-environmental transformations interact with broader political and economic dynamics as well as how people imagine and perceive such transformations. She holds a PhD in environmental science and technology (ICTA-Universitat Autònoma de Barcelona), an MSc in human geography (Universiteit van Amsterdam), and a BSc in political science (Universidad Complutense de Madrid).

Kaitlin Strange is a researcher and PhD candidate at the Laboratory of Urban Transformation and Global Change (TURBA) at IN3-Universitat Oberta de Catalunya (UOC). Her research focuses on equitable urban climate adaptation, grassroots organisations, and the ways in which cities integrate transformative principles into adaptation planning and implementation. Kaitlin has worked for more than 10 years as an urban planner and consultant working with nonprofits and government agencies alike. She is committed to community and data-driven planning to address just and sustainable futures.

Erik Swyngedouw is Professor of Geography at the University of Manchester, UK. He was previously Professor of Geography at Oxford University and held the Vincent Wright Visiting Professorship at Science Po, Paris, 2014. He also holds honorary doctorates from Roskilde University and the University of Malmö. His research focuses on political ecology, environmental politics, democratisation, urbanisation, politicisation, and socio-ecological movements. He is author of *Urban Political Ecology in the Anthropo-Obscene* (edited with Dr. H. Ernstson, Routledge), *Promises of the Political* (MIT Press), *Liquid Power: Contested Hydro-Modernities in twentieth Century Spain* (MIT Press), and *Social Power and the Urbanization of Nature* (Oxford University Press).

Christos Zografos is Ramón y Cajal Senior Research Fellow at Johns Hopkins University – Universitat Pompeu Fabra (JHU-UPF) Public Policy Centre in Barcelona, Spain. He is also vice director of the research group Health Inequalities, Environment, and Employment Conditions Network in the Department of Political and Social Sciences at Universitat Pompeu Fabra. Christos is a critical environmental social scientist working in the fields of political ecology, social ecological economics, and degrowth. He has published widely on climate policy implementation conflicts, value plurality and sustainability decision-making, direct democracy and degrowth, and social narratives towards socio-environmental transformation. He holds degrees from the University of Edinburgh, Universitat Autònoma de Barcelona, and Athens University of Economics and Business.

Part I
The Political Ecology of Urban Resilience

Chapter 1
Resilience for All or for Some? Reflections Through the Lens of Urban Political Ecology

Hug March and Erik Swyngedouw

Abstract In this first chapter of the book, we develop a critical perspective of urban resilience through the lens of urban political ecology, with an eye towards charting a trajectory that may open new political possibilities. The chapter is divided into five parts. The first part demonstrates how the urban and the urbanisation process implies an uneven distribution of risks and vulnerabilities. In the second part, we shall explore how the notion of *resilience* should be understood as a *scientific* concept dressed in an ideological mask that has begun to be problematised by critical scholars. In the third part, we discuss how political ecology may contribute to the emerging critique of urban resilience, and in the fourth part, we develop how urban resilience operates as an immuno-biopolitical fantasy. The final section explores ways through which the urban socio-ecological condition can be repoliticised, opening new possibilities for a more democratic and progressive urban resilience decision-making. By doing so, we suggest a research agenda for transformative, more democratic, emancipatory, and socially grounded forms of urban resilience.

Keywords Urban political ecology · Planetary urbanism · Urban resilience · Biopolitics · Climate adaptation

---

H. March (✉)
Estudis d'Economia i Empresa and Internet Interdisciplinary Institute (IN3), Universitat Oberta de Catalunya (UOC), Barcelona, Spain
e-mail: hmarch@uoc.edu

E. Swyngedouw
School of Environment, Education and Development, Department of Geography, The University of Manchester, Manchester, UK

I. Ruiz-Mallén et al. (eds.), *Urban Resilience to the Climate Emergency*, The Urban Book Series, https://doi.org/10.1007/978-3-031-07301-4_1

1.1 The Twenty-First Century: The Age of Planetary Cities of Extremes

> Stop calling me resilient. I'm not resilient. Because every time you say, 'Oh, they're resilient', you can do something else to me. (Tracie Washington, New Orleans-based civil rights attorney 2010)

In 2005, Hurricane Katrina exposed in the words of political ecologist Paul Robbins the *structural inequalities of race and class, which are physically inscribed into the seascape, implicated in the ecological transformation of the coastal zone, and inseparably linked to the technologies that govern the flow of water through the Mississippi delta* (Robbins 2012, p. vii). Ten years after this disaster, the Rockefeller Foundation, which has played a fundamental role in the popularisation and implementation of urban resilience in many cities through its 100 Resilient cities initiative, published a post called *New Orleans & The Birth of Urban resilience.*[1] This post pointed out how, after a decade of interventions focused on building resilience -and where this organisation was a key actor – *New Orleans has rightly been called a success story of building urban resilience.*

Far from those celebratory narratives on urban resilience in New Orleans and reflecting on Tracie Washington's clarion call to *Stop Calling me resilient*, Maria Kaika (2017a) suggests taking her claim seriously. What Tracie calls for is to focus on the things and processes that create the need for her to be resilient in the first place. Living in New Orleans, and surviving both Hurricane Katrina and the BP oil spill in the Gulf of Mexico, Tracie's comment reflects on the deeply racialised, gendered, and class-based character of the alleged *resilient* recovery of New Orleans post-Katrina. Urban resilience seems to take a form whereby eco-gentrification paves the way – and not surprisingly so – to a more resilient urbanity.

The counterintuitive claim of *Stop calling me resilient* serves to point to the antinomies of the resilience argument and situates the question of resilience within the contours of present-day urbanisation. Not surprisingly, urban *resilience* is high on the political and academic agenda, although not free of contradictions (Elmqvist et al. 2019; Meerow and Neuner 2021). Many cities, especially in the Global North, are positioning themselves as important actors to tackle the climate emergency, thereby demonstrating how cities have shifted from being passive, yet responsible, actors within the climate (and more generally environmental) conundrum to active agents (at least discursively) in tackling the global environmental crisis (Satorras et al. 2020). Cities are both key drivers in the process of combined and uneven socio-ecological transformation and face serious, albeit socially highly triaged, consequences of accelerating socio-ecological disintegration (Angelo and Wachsmuth 2020). This position must be fully endorsed and confronted in any theoretical discussion or practical interventions that articulate the question of resilience. In times of climate emergency and in a context of planetary urbanisation (Brenner 2018),

[1] See https://www.rockefellerfoundation.org/new-orleans-birth-urban-resilience/ – accessed 9 February 2022.

where recurrent socio-environmental disasters, injustices, inequalities, and uneven socio-ecological disintegration are experienced in the urban fabric in both Global North and Global South cities, *it is urgent to think about and engage in emancipatory and political socio-ecological transformations* (Ernstson and Swyngedouw 2019, p. 3). In the face of these uneven development trends, increasing inequality and power imbalances, it is vital to problematise mainstream socio-environmental discourses and visions that tend to depoliticise the nature of environmental problems, reducing them to technical, managerial, and scientific issues while presenting an undifferentiated humanity that faces those problems.

Almost 20 years ago, one of the key figures in critical geography and social sciences, David Harvey, argued that *[e]cological arguments are never socially neutral any more than socio-political arguments are ecologically neutral. Looking more closely at the way ecology and politics interrelate then becomes imperative if we are to get a better handle on how to approach environmental/ecological questions* (Harvey 1993, p. 25). Understanding the way policymakers, scientists, or practitioners analyse, problematise, and prefigure solutions for the climate emergency through *urban resilience* narratives (Meerow and Neuner 2021), which are deeply infused by worldviews and values, and traversed by power relations, is critical to deflect the socio-ecological history of the future in more egalitarian and socially just directions.

The urbanisation process and urban life more generally are shaped by extraordinary and deepening socio-ecological polarisation and forms of exclusion, both within and between cities (Dawson 2017). The combined and uneven socio-ecological dynamics that mark the contemporary urbanisation process demonstrate both the mechanisms that expose a range of people to increasing risk and symptomatically illustrate the highly triaged and unequal exposure of people to socio-ecological risk and uncertainty.

The urbanisation process and *sustaining* urban life globally is indeed predicated on excessive resource appropriation and transformation, often in the most vulnerable places on earth. Consider, for example, how the urban Information Technology apparatus mobilises all manner of resources, such as Coltan, Lithium, Copper, or rare earth elements, thereby sustaining an extractivist socio-ecological regime with its socio-ecological devastations and increasing vulnerability for those dispossessed from their land or exposed to the hazards that accompany resources extraction (e.g., March 2018; Swyngedouw 2022a). Therefore, the resilience of some places to socio-ecological shocks is often bought at the cost of increasing vulnerability and exposure in some other places and for other disenfranchised groups. In sum, cities are central nodes in sustaining the process of uneven socio-ecological transformation (Kaika and Swyngedouw 2014). Indeed, the urbanisation process intensifies what Marx and others called a *metabolic rift* (see Foster 2000), that is, the de-territorialisation, socio-metabolic transformation, and re-territorialisation of all manner of non-human material, thereby structuring a kaleidoscopic landscape of differential vulnerabilities, capabilities, and risks (Swyngedouw 2006; Arboleda 2020). The urbanisation process demonstrates how resilience is not only unevenly distributed but fundamentally produced by the various dynamics that underpin the urbanisation process.

Finally, communities within cities are also prone to all manner of risks with uneven capacities to deal with socio-ecological shocks. As the example of New Orleans post-Katrina demonstrates, some social groups exhibit a much greater capacity to absorb shocks than others do. Whether we consider earthquakes, exposure to toxins, floods, heat waves or the spread of disease, resilience is primarily the concern of the poor and disempowered. The recent COVID-19 pandemic and its socio-ecological triaging demonstrated how the poor generally carried a much greater risk than the elite in suffering from the pandemic. For example, in the U.K., BAME (Black, Asian and other minorities) groups were up to four times more likely to become infected and suffer death from COVID-19 compared with whites (Otu et al. 2020). Urban resilience, therefore, is not a disembodied and homogenous process but one that is uneven and shaped by the unequal exposure to risk as structured by the process of urban socio-ecological change. This turns the question of resilience into a political and politicising terrain. It is from this perspective that we consider the question of *urban resilience* in this chapter.

1.2 From Mainstreaming Urban Resilience to the Birth of a Critical Scholarship of the Politics of Resilience

The concept of resilience has a long and established history and has been articulated in a range of academic fields, from psychology, through engineering or disaster risk reduction, to the analysis of socio-ecological systems (Gunderson 2000; Matyas and Pelling 2015). Alexander (2013) traces the historical evolution of the concept of resilience, arguing that the notion of *resilio* or *resiliere* was present in standard Latin (meaning *bounce*) and appears in classical texts. Alexander (2013) contends that Francis Bacon was *responsible for the first known scientific use in English of the word 'resilience'* (p. 2709) in his natural history compendium *Sylva Sylvarum*. Since the mid-nineteenth century, the notion of resiliency *was used in the sense of the ability to withstand the effects of earthquake*[s] (p. 2709). Around that time, mechanical engineering also began to use the notion of resilience. One century later, in the 1950s, psychology started using the term and popularised it in the final decades of the twentieth century. It would be Crawford Stanley (Buzz) Holling who would bring the concept to ecological sciences (1973), and it was further developed through the work of the Stockholm Resilience Centre.[2] By the late twentieth century, the notion of resilience also moved to the study of human ecology while expanding the very meaning inscribed in the concept to include social resilience.

Contemporary socio-ecological resilience approaches emerged in an optimistic turn by foregrounding the formidable capacities that humans and forms of social

[2] see https://www.stockholmresilience.org/

organisation possess to deal with problems related to global environmental change. From the early writings of Holling to the re-emergence of the interest in the concept in the early twenty-first century (Gunderson 2000; Carpenter et al. 2001; Folke et al. 2002), resilience has been consolidated as a scientific concept that is concerned with change, dynamics, and uncertainties of complex socio-ecological systems. In this regard, a widespread understanding of resilience relates to the capacity of a system to absorb disturbance and reorganise while undergoing change but still retaining essentially the same function, structure, identity, and feedback loops (Walker et al. 2004). However, such capacities will not necessarily translate into action, raising critical questions as to why, how, to whom, and where they make an impact on real adaptation (to climate change or other socio-environmental hazards) and transformation (Pike et al. 2010; Berrang-Ford et al. 2011; Nelson 2011; Meerow et al. 2016).

The metaphor of urban resilience has evolved in a context of rapidly deepening and complex challenges galvanised by the urbanisation process (Galderisi and Colucci 2018). It has emerged as a buzzword alongside other concepts such as the sustainable city or the smart city (de Jong et al. 2015; March 2022). Early approaches to urban resilience considered how a resilient city could withstand an intense shock event without the outbreak of chaos, disintegration, or permanent deformation. Therefore, those approaches were designed to permit the urban environment to anticipate, endure, and recover from the effects of natural or technological hazards (Godschalk 2003). Since the second decade of the twenty-first century, however, and coinciding with the renewed momentum of the role of cities in shaping the global socio-environmental crisis, and more specifically, the climate emergency (Satorras et al. 2020), a plethora of new definitions of urban resilience has permeated academic and policy debates and practice (Meerow and Neuner 2021). According to the state-of-the-art review by Meerow et al. (2016), conceptualisations of urban resilience differ concerning their characterisation of the *urban*, the notion of equilibrium, their normative perspective, the possible pathways to resilience, the understanding of adaptation, and the timescale of action.

In sum, urban resilience has gained popularity among academic and policy-making circles in the past few years in light of the international discussions on and concerns with how cities could tackle the effects of the climate emergency. For example, the organisation *Local Governments for Sustainability* (ICLEI) (2018) considers a resilient development pathway one of the five critical pathways for achieving a sustainable world. Similarly, UN-related agencies and programmes have promoted the idea of resilient cities through SDG 11 (Sustainable Cities and Communities) of the 2030 Agenda for Sustainable Development, the UNISDR campaign of *Making cities resilient!*, or the City Resilience Profiling Programme (CRPP) developed by UN-Habitat.

Indeed, the concept of urban resilience has captured both academic – through grants and projects – and policy makers' attention – through establishing strategies of urban resilience, partnering with companies/universities to experiment with resilience solutions, or joining global initiatives such as the 100 Resilient Cities organisation (funded by the Rockefeller Foundation), which later evolved into the Resilient Cities Network (Bliss 2019).

In light of the overtly over-celebratory and optimistic versions of resiliency, critical scholars have scrutinised the concept and practice during the past few years, opening a critical research agenda through a series of crucial arguments. The first argument, arguably, revolves around the imperious need to scrutinise how the notion is discursively built and articulated in various geographical settings. In 2012, Porter and Davoudi (2012) and Davoudi et al. (2012) wrote a *cautionary note* on the politics of resilience for planning, underscoring that it was not clear yet what resilience meant by that time. Along similar lines, MacKinnon and Derickson (2012, p. 254) argue that *the frequency by which resilience is invoked by progressive activists and movements underlines the need for critical appraisal of both the term itself and the politics it animates*. MacKinnon and Derickson argue that, to some extent, resilience has been constitutive of the development of capitalism in the twenty-first century: *resilient spaces are precisely what capitalism needs – spaces that are periodically reinvented to meet the changing demands of capital accumulation in an increasingly globalized economy* (p. 254). Walker and Cooper (2010, p. 144, cited in MacKinnon and Derickson 2012, p. 254) situate resilience as *a pervasive idiom of global governance* which is *abstract and malleable enough to encompass the worlds of high finance, defence and urban infrastructure*. In this sense, *resilience* chimes well with a culture of (neo)liberal individualism that places the onus on the individual or community to build internal defence structures against a presumably *external* threat while eschewing concerns of solidarity or public intervention.

Urban resilience might be *old wine in new bottles* (Kaika 2017a), reproducing apolitical ecological views that are nonetheless shaped by uneven power choreographies, inequalities, and injustices (MacKinnon and Derickson 2012). This echoes what Peck and Tickell (2002) framed as roll-out neoliberalism to describe new technologies of governments and new modes of governance that deepen further the culture of neoliberalisation understood as individual or community self-responsibilisation. With respect to resilience planning, this implies maintaining the status quo and reinforcing the neoliberal fantasy whereby resilience is the domain and responsibility of individual action and local community organisation in the face of external *risk* (Saurí 2018). Resilient communities, therefore, could also be sidelined in processes of collective solidarity and responsibility. Although not explicitly from a political ecology perspective, Meerow et al. (2016) emphasise the power relations that shape the meaning of urban resilience: *who determines what is desirable for an urban system? Whose resilience is prioritized? Who is included (and excluded) from urban systems*? Or, *what are the underlying motivations for building urban resilience* (p. 46). Briefly, urban resilience needs to be considered in the image of what, for whom, why, as well as when and where (see also Chap. 5 for the specific case of Latin America).

In another dimension, critical scholarship has pointed out the lack of understanding of the multiple and interlinked sources of vulnerability, emphasising the need to focus on structural vulnerabilities (Wilbanks and Kates 2010), and thereby avoiding centring resilience on a single stressor (e.g., water, heat) (Ajibade and McBean 2014; Harrison and Chiroro 2017). This again resonates with the question of *resilience to what* and *where*, pointing out the importance of considering uneven development. Urban resilience in the Global North may have very different meanings than urban resilience in the Global South, not to mention the vast array of contingencies and pre-existing vulnerability structures different urban geographies may face. It is, therefore, vital to scrutinise the uneven socio-spatial patterning of urban resilience (Sapountzaki 2007; Pike et al. 2010; Suárez et al. 2016). For instance, Elliott and Pais (2006) and Elliott and Sams-Abiodun (2010) focus on the uneven response (and resilience) to hurricane Katrina in New Orleans, as we already pointed out in the introduction to this chapter. Elsewhere, in Barcelona, Marí-Dell'Olmo et al. (2018) have shown that vulnerability to heat-related events is shaped by social and economic characteristics, posing new challenges to designing socially sensible resilience strategies within the city. Therefore, questions such as *whose resilience is being addressed* have been raised, acknowledging the uneven socio-spatial configuration of urban environments (Vale 2014; see also Meerow et al. 2016).

Finally, we should point out the challenges behind building diverse and inclusive communities active in resilient transformations. Along these lines, scholars have identified contradictory directions between the theory and practice of urban resilience (e.g., Stumpp 2013), especially in what concerns climate adaptation, resulting in socially unjust outcomes (Anguelovski et al. 2016, 2019; Chu et al. 2017; see also Chap. 2 for an in-depth engagement with environmental justice scholarship). Shedding light on the uneven impact of urban resilience while including equity issues in the resilience agenda has long been a demand of those advocating a more socially grounded perspective on resilience (Leichenko 2011; Brown 2014; Matin et al. 2018; Meerow et al. 2019; Mees et al. 2019).

In light of this growing body of critical scholarship that calls for a more situated and nuanced understanding of the politics of resilience, we aim to contribute further to such a critique through the lens of Urban Political Ecology, discussing mainstream resilience approaches as an immune-biopolitical fantasy.

1.3 Urban Political Ecology As a Lens to Contribute to Critical Scholarship on Urban Resilience

Urban political ecology (UPE) focuses on the socio-ecological inequalities embodied in and shaped by the production and reproduction of capitalist urbanisation itself (Keil 2003, 2005; Heynen et al. 2006). The theoretical objective of UPE is to explore the process of socio-natural assembling through which urban landscapes become constructed as a hybrid enmeshing of both humans and non-human objects and pivoting around the social inequalities that are expressed in and by urban socio-ecological metabolic processes (Swyngedouw and Heynen 2003). Those metabolic processes render the urbanisation process a highly unequal socio-ecological configuration whose functions (e.g., sustainability, resilience, smartness, competitiveness) are predicated upon geographically and ecologically widening networks of socio-ecological transformation (Heynen et al. 2006; Swyngedouw 2006). In the present context, sustaining urbanisation requires socio-ecological relations and networks (from the extraction of materials and food production to the transformation of energy) that are planetary (Arboleda 2016). Indeed, large scale or *planetary* urbanisation (Brenner 2018) of all manner of natures constitutes *the* spatial form of capital accumulation with all sorts of intended and non-intended, but thoroughly unequal, outcomes. The uneven positions that different social groups take within these circuits are symptoms expressing the dynamics of the underlying processes of socio-ecological metabolic circulation.

The production of urban environments and the *metabolic vehicles* that assure its functioning -such as infrastructures of all kinds, the technical and institutional conditions that permit the flow and metabolisation of energy, water, food, information, bodies, and things as well as their socio-ecological characteristics- are of course mediated by governing arrangements that are often nominally democratic, but are nonetheless necessarily profoundly committed to ensuring the uninterrupted expansion of the capital circulation process. It is precisely this articulation between state, class, and environmental translation that renders urban socio-ecological processes, including the question of *sustainability*, highly conflictive and subject to intense political and social struggle. This process also exposes individuals and social groups differentially to all manner of anticipated or non-anticipated risks, dangers, and/or vulnerabilities. From this political-ecological perspective, urban socio-ecological conditions and the configurations of their governance are never just local but are attached to processes that operate in and through the widely variegated and diverse ecologies of the world. Such urban political-ecological approaches highlight the political core of environmental change and transformations and insist on the fundamentally political nature of the modes of socio-technically organising the metabolic transformation of nature.

Therefore, UPE is concerned with the democratic political processes through which socio-ecological transformations occur. Thus, rather than invoking a normative notion of environmental justice, political ecology insists on focusing on the realities of the presumed democratic political equality in the decision-making processes that organise socio-ecological transformation and choreograph the management of the commons. In doing so, attention shifts from a techno-managerial to a resolutely political vantage point (articulated around the notion of socio-ecological and political equality) that considers the ecological conundrum to be inexorably associated with democratic political action (Kaika and Swyngedouw 2014).

In sum, UPE challenges apolitical views about socio-ecological relations (Robbins 2012), aiming *to illustrate the myriad articulations of how urban environmental and social change co-determine each other* (Heynen 2014, p. 598). Through these lenses, the *urban* is presented as a socio-natural constellation that encapsulates and articulates the power choreographies and relations through which it is constantly being made and remade (Gandy 2002; Heynen et al. 2006; Loftus 2012).

Along these lines, we can conceptualise urban resilience strategies as a set of complex socio-ecological, technological, political, and economic processes, which are not only infused by, but also, reshape social power relations. Resilience, from this perspective, is not about indexing and reorganising the capacity of individuals or social groups to counter adverse effects generated by external intruders that threaten to undermine or destabilise the integrity of the body or the community. On the contrary, it is the internally constituted socio-ecological relations that define relative places of vulnerability and resilience. Improving the socio-ecological integrity of the body or community resides primarily in altering the relational configuration that produces the unevenly triaged urban landscapes of vulnerability.

In a sense, we can argue that mainstream resilience might be driven by an eco-prophylactic objective, whereby techno-managerial adaptations and socio-institutional resilience are promoted to ensure nothing really has to change, i.e., hegemonic socio-ecological relations can be *sustained* for a while longer without risk of disintegration of vulnerable livelihoods. This is precisely what we turn to in the next section.

1.4 Urban Resilience As an Immuno-Biopolitical Fantasy

In light of the above critical interventions, the signifier *resilience* refers to a potentially almost infinite set of signified, rendering the concept relatively empty in terms of content or, in other words, its content can be filled with a wide range of competing and occasionally contradictory meanings. It is precisely this *emptiness* that turns *resilience* into a flexible fantasy notion that exerts an irresistible ideological pull. Here perhaps resides the key attraction and power of *resilience* as a metaphor and diverse set of practices. Together with *sustainability*, *adaptation*, and *mitigation*, *resilience* arguments have become dominant discourses in the debate over socio-ecological transformations.

The phantasmagorical promise of urban resilience provides a seemingly immunological prophylactic against the threats and dangers posed by the often unpredictable and complex acting of an assumedly external Nature. The key objective of much of urban sustainability, *resilience*, *smart* technologies and adaptive eco-managerial policies and practices that dominate urban ecological policies and interventions are indeed precisely aimed at re-enforcing the immunological capabilities of the individual or social *immune* system against dangerous or threatening intruding things and processes *so that* life as we know it can continue. Is this not an extension of the process of cordoning off and sequestration when infectious diseases threaten to spatialise in manners that could harm or negatively influence the immuno-engineered *sustainable* or *resilient* bubbles of the elite's local life (see also Kaika 2017a; Swyngedouw and Ernstson 2018)?

Resilience, therefore, adds to the arsenal of discourses and practices that Roberto Esposito (2011) identifies as constituting an integral part of what he calls a neoliberal immuno-biopolitical governance regime. With this conceptualisation, Esposito expands on Michel Foucault's notion of biopolitical governmentality. While Foucault insisted on the processes by which states became concerned with the physical and social health of the population, Esposito argues that, as a result of globalisation and the destabilising threats emanating from this process, biopolitical governance has now been extended to include an immunological desire. The latter refers to the increasing concern of governments with protecting objects of government (the population) from possibly harmful intruders and threatening, *risky*, or destabilising outsiders that might undermine the integrity and socio-ecological coherence, if not sheer survival, of the population. In doing so, immuno-biopolitics aims at assuring that life-as-we-know-it can continue to be lived (Esposito 2011).

Immunological, in this context, refers to the suspension of the obligation of communal gift-giving. It constitutes a form of asylum that suspends the obligation to be concerned with the commons or with the polity beyond the confines of the individual body or local community. The (neo)liberal credo to enjoy individual freedom and choice is precisely the founding gesture of such an immunological bio-politics, i.e. the relative isolation or sequestration of the body from its insertion in the obligations and violence that bonds (global) community life (Swyngedouw and Ernstson 2018). Alain Brossat (2014) refers to this as a fantasy of *immunitary democracy*. He argues that the immunitary dispositive runs the danger of destroying a sense of community and, therefore, inaugurates the end of politics. Indeed, the immuno-biopolitical frame produces simultaneously the exposed and the exiled (the non-immunised) as the flipside of the immunised body. The immuno-biopolitical gesture turns into a necropolitics for the exposed. In other words, *resilience* for some implies vulnerability for others. This is clearly illustrated by MacKinnon and Derickson (2012, p. 254) when they argue that *the resilience of capitalism is achieved at the expense of certain social groups and regions that bear the costs of periodic waves of adaptation and restructuring* (p. 254). As Swyngedouw (2022b, p. 63), building on Kaika (2017b), puts it, *such immunological sanctity space offers only the affective politics of either hate or compassion for the threatening intruder as flipsides of the same coin, while sustaining their expulsion into the peripheral zones of socio-ecological disintegration and enclaves of poverty where life remains bare*. It is precisely the climate emergency that provokes extreme vulnerability for some, while others can sustain their lives in zones of resilience.

As Roberto Esposito (2011) argues further, the immunological biopolitical drive turns indeed into a thanatopolitics or a necropolitics, centring on the question of *who should live and who should die*. In the excessive acting of the immunological drive, the dispositive turns against what it should protect. It becomes self-destructive in a process of auto-immunisation. The construction of resilient and sustainable urban enclaves for the privileged is paralleled by the making of the unprotected exiles and by intensifying socio-ecological disintegration elsewhere. This is eco-gentrification at its best. In other words, the mechanisms that promise to secure the future of life in some places end up undermining life elsewhere, at all geographical scales. Becoming *resilient* is indeed the ultimate hysterical act that promises to protect the self, often at a cost to the Other. The only sane response to this is indeed *Stop calling me resilient!* (Kaika 2017a). This phantasmagoric staging of resilience as panacea depoliticises the matter of nature. We can survive and do so without the necessity of facing political actions and radically different political choices. Therefore, the critical challenge is to recover and foreground the political substance of resilience. Furthermore, this requires both to cut through the fantasy that underpins the indelible lure of resilience and to chart trajectories for a politicising conceptualisation and practice of resilience.

1.5 Politicising Urban Resilience: Traversing the Fantasy

Notwithstanding the above critical account of mainstream notions of urban resilience as an immuno-biopolitical fantasy that depoliticises nature, and in light of the nascent critical scholarship on urban resilience pointed out in the second section of the chapter, we argue that there is still a potential transformative power in considering and pursuing alternatives modes of urban resilience (see Chap. 2 for a review of resilience and transformation and Chap. 10 for a synthesis of the limitations and potentials of institutional and grassroots climate resilience initiatives to build transformational processes). To move towards a politically engaged form of resilience, more attention should be paid to social and distributive matters, addressing questions such as whose resilience or who counts for the governance of urban resilience. Although we can identify in the early 2000s attempts to link political ecology with ecological resilience (see, for instance, Peterson 2000), the terrain has been more productive in the past few years in light of increasing critical research on resilience. For example, Matthew Turner (2014) interrogates the possibilities of an alliance between political ecology and resilience thinking, showing some common features in the intellectual trajectories of both fields and pointing out significant divergences in terms of normative commitments. Ingalls and Stedman (2016) further contribute to finding shared intellectual spaces between critical approaches to resilience and political ecology. They depart from the premise that power relations have been systematically neglected (or at most played a very marginal role) by mainstream resilience frameworks and approaches.

It is crucial to discursively foreground and excavate the power imbalances that infuse resilience practices if a more grounded and emancipatory urban resilience is to be pursued and implemented. The inclusive engagement of stakeholders remains a crucial challenge to democratise and even politicise urban resilience decision-making (a question that is centrally taken up in Chaps. 3, 4, 5, 6, 7, 8, and 9). Alternatives to top-down governance include participatory/collaborative forms of governance such as, for example, the co-production of resilience (Borquez et al. 2016; Fagan-Watson and Burchell 2016; Muñoz-Erickson et al. 2017) and bottom-up initiatives such as neighbourhood resilience (Stevenson and Petrescu 2016) and community resilience practices (Norris et al. 2008; Magis 2010; Ruiz-Mallén and Corbera 2013; Delgado-Serrano et al. 2018).

MacKinnon and Derickson (2012, p. 254), through their work with grassroots movements in Glasgow and more generally in Britain, recognise that the notion of resilience has helped *to frame particular forms of activism, some of which are anti-capitalist in nature*. Some scholars argue that resilience may enable transformative avenues to address socio-environmental problems when grassroots movements lead those resilience plans or strategies or when community groups are deeply involved (see Camps-Calvet et al. 2015; Bródy et al. 2018). Along those lines, Satorras et al. (2020) show the possibilities that the co-production of urban sustainability policies, including urban resilience, may bring to transformative adaptation. In other words, while there is the risk to be co-opted by mainstream resilience approaches, we still discern possibilities within politically engaged and progressive forms of resilience that shift focus towards questions articulated around equality and socio-ecological distribution.

The call for a more politicised approach to the question of resilience requires, therefore, several shifts in the discursive and praxis-based approaches invoked under the banner of *resilience*. The anthropomorphism that transfigured the notion of resilience from a particular natural science view of ecosystems development to include human-nature relations has to be rejected entirely. Indeed, the naturalisation of complexity theory and the dynamic autopoietic self-organisation of panarchic systems that ecologists and other natural scientists uphold as the key model to explain ecological stability and transformations cannot and should not be extended uncritically to the human world. While ecological *resilience* marks the intrinsic capacity of a complex ecological configuration to withstand shock and crisis, the move to also considering socio-ecological systems takes the problem from the domain of *natural* dynamics to an assemblage in which social power, political positions, and individual preferences play a crucial role. Anthropomorphising resilience invariably leads to a naturalisation of the social and its particular embedding within socio-ecological processes and disavows the historically and geographically contingent socio-ecological power relations that shape differential vulnerabilities, capacities, and exposure to risks.

Much of the natural science perspective on resilience is based on the consolidation of non-linear ecological complexity theory that centres on *emergence*, *resilience*, continuous experimentation, the *indeterminacy* of nature, and radical openness (Holling 1973; Folke 2006), but does not pay explicit attention to social power relations, differential cultures, social change, and exclusions/inclusions choreographed by governance arrangements (see Nadasdy 2007; Hornborg 2009; Cote and Nightingale 2012). As Bruce Braun (2015) argues, drawing on Sara Nelson (2015), we need to acknowledge the historically parallel ways through which non-deterministic geosciences (including complexity science and resilience theory) emerged in the 1970s alongside the rising prominence of the phantasmagorical promise of neoliberalism (see Protevi 2013; Pellizzoni 2015). Both rose to prominence during the structural crisis of capitalism in the 1970s and the parallel attempts to search for a fix to the subsequent socio-economic decline (see Walker and Cooper 2010; Nelson 2015). Indeed, complexity theory and related perspectives have played an essential part in the rise to prominence of neoliberal socio-ecological relations. This, in turn, permitted approaching *resilience* as an assemblage of activities (and associated actors) that held the promise of the existing socio-ecological frame to continue as communities become more *resilient* against the successive and inevitable shocks and crises produced by the combined and uneven development of the neoliberal socio-ecological order.

Foregrounding the politicisation of resilience requires, therefore, re-centring the unequal power relations and socio-ecological positions in the relational networks of the socio-ecological constellation, with an eye towards identifying how political equality and democratisation can be enhanced. Enhancing *resilience*, therefore, necessitates taking sides with those who are most vulnerable and re-structuring socio-ecological relations such that vulnerabilities and risks are not only lowered but also distributed more widely and evenly. The politicisation of resilience requires politicising urban movements that seek to transform both procedures of environmental governance and the unequal socio-ecological relations that sustain present-day urbanisation. Considering the possibilities and potentialities of such moves is precisely what this book seeks to explore.

Acknowledgements We are grateful to the co-editors of this book, I. Ruiz-Mallén and M. Satorras, for their comments and editing provided to earlier versions of this manuscript. We also thank Ramon Ribera-Fumaz for his insightful suggestions. This research has been funded by the Spanish Research Agency (Agencia Estatal de Investigación, AEI) through the Plan Nacional RESCITIES (PGC2018-100996-AI00(MCIU/AEI/FEDER, UE)).

References

Ajibade I, McBean G (2014) Climate extremes and housing rights: a political ecology of impacts, early warning and adaptation constraints in Lagos slum communities. Geoforum 55:76–86

Alexander DE (2013) Resilience and disaster risk reduction: an etymological journey. Nat Hazards Earth Syst Sci 13:2707–2716

Angelo H, Wachsmuth D (2020) Why does everyone think cities can save the planet. Urban Stud 57(11):2201–2221

Anguelovski I, Shi L, Chu E, Gallagher D, Goh K, Lamb Z, Reeve K, Teicher H (2016) Equity impacts of urban land use planning for climate adaptation: critical perspectives from the global north and south. J Plan Educ Res 36(3):333–348

Anguelovski I, Connolly JJT, Pearsall H, Shokry G, Checker M, Maantay J, Gould K, Lewis T, Maroko A, Roberts JT (2019) Why green "climate gentrification" threatens poor and vulnerable populations. PNAS 116(52):26139–26143

Arboleda M (2016) In the nature of the non-city: expanded infrastructural networks and the political ecology of planetary urbanisation. Antipode 46(2):233–251

Arboleda M (2020) Planetary mine: territories of extraction under late capitalism. Verso, London

Berrang-Ford L, Ford JD, Paterson J (2011) Are we adapting to climate change? Glob Environ Chang 21(1):25–33

Bliss L (2019) The rise, fall, and possible rebirth of 100 Resilient Cities CityLab, 12 June 2019

Borquez R, Aldunce P, Adler C (2016) Resilience to climate change: from theory to practice through co-production of knowledge in Chile. Sustain Sci 12:163–176

Braun B (2015) New materialisms and neoliberal natures. Antipode 47(1):1–14

Brenner N (2018) Debating planetary urbanization: for an engaged pluralism. Environ Plann D 36(3):570–590

Bródy LS, Chelleri L, Baró F, Ruiz-Mallén I (2018) Enhancing community resilience in Barcelona: addressing climate change and social justice through spaces of co-management. In: Galderisi A, Colucci A (eds) Smart, resilient and transition cities: emerging approaches and tools for a climate-sensitive urban development. Elsevier, pp 203–208

Brossat A (2014) La démocratie immunitaire. La Dispute, Paris

Brown K (2014) Global environmental change I: a social turn for resilience? Prog Hum Geogr 38(1):107–117

Camps-Calvet M, Langemeyer J, Calvet-Mir L, Gómez-Baggethun E, March H (2015) Sowing resilience and contestation in times of crises: the case of urban gardening movements in Barcelona. Partecipazione e Conflitto 8(2):417–442

Carpenter S, Walker B, Anderies JM, Abel N (2001) From metaphor to measurement: resilience of what to what? Ecosystems 4(8):765–781

Chu E, Anguelovski I, Roberts D (2017) Climate adaptation as strategic urbanism: assessing opportunities and uncertainties for equity and inclusive development in cities. Cities 60:378–387

Cote M, Nightingale AJ (2012) Resilience thinking meets social theory: situating social change in socio-ecological systems (SES) research. Prog Hum Geogr 36(4):475–489

Davoudi S et al (2012) Resilience: a bridging concept or a dead end? The politics of resilience for planning: a cautionary note. Plan Theory Pract 13(2):299–333

Dawson A (2017) Extreme cities – the peril and promise of urban life in the age of climate change. Verso, London

de Jong M, Joss S, Schraven D, Zhan C, Weijnen M (2015) Sustainable-smart-resilient-low carbon eco-knowledge cities; making sense of a multitude of concepts promoting sustainable urbanization. J Clean Prod 109:25–38

Delgado-Serrano M, Oteros-Rozas E, Ruiz-Mallén I, Calvo-Boyero D, Ortiz-Guerrero C, Escalante-Semerena RI, Corbera E (2018) Influence of community-based natural resource management in the resilience of social-ecological systems. Reg Environ Chang 18(2):581–592

Elliott JR, Pais J (2006) Race, class, and Hurricane Katrina: social differences in human responses to disaster. Soc Sci Res 35:295–321

Elliott JR, Haney TJ, Sams-Abiodun P (2010) Limits to social capital: comparing network assistance in two New Orleans neighborhoods devastated by Hurricane Katrina. Sociol Q 51(4):624–648

Elmqvist T, Andersson E, Frantzeskaki N et al (2019) Sustainability and resilience for transformation in the urban century. Nat Sustain 2(1):267–273

Ernstson H, Swyngedouw E (2019) Politicizing the environment in the urban century. In: Ernstson H, Swyngedouw E (eds) Urban political ecology in the Anthropo-Obscene – interruptions and possibilities. Routledge, London, pp 3–12

Esposito R (2011) Immunitas. Polity Press, Cambridge

Fagan-Watson B, Burchell K (2016) Heatwave planning: community involvement in co-producing resilience. Build Res Inf 44(7):754–763

Folke C (2006) Resilience: the emergence of a perspective for social–ecological systems analyses. Glob Environ Chang 16(3):253–267

Folke C, Carpenter S, Elmqvist T, Gunderson L, Holling CS, Walker B (2002) Resilience and sustainable development: building adaptive capacity in a world of transformations. Ambio 31(5):437–440

Foster JB (2000) Marx's ecology: materialism and nature. Monthly Review Press, New York

Galderisi A, Colucci A (2018) Smart, resilient, and transition cities: commonalities, peculiarities and hints for future approaches. In: Galderisi A, Colucci A (eds) Smart, resilient, and transition cities. Emerging approaches and tools for a climate-sensitive urban development. Elsevier, Cambridge

Gandy M (2002) Concrete and clay: reworking nature in New York City. MIT Press, Cambridge, MA

Godschalk DR (2003) Urban hazard mitigation: creating resilient cities. Nat Hazard Rev 4(3):136–143

Gunderson LH (2000) Ecological resilience—in theory and application. Annu Rev Ecol Syst 31:425–439

Harrison E, Chiroro C (2017) Differentiated legitimacy, differentiated resilience: beyond the natural in 'natural disasters'. J Peasant Stud 44(5):1022–1042

Harvey D (1993) The nature of environment: the dialectics of social and environmental change. In: Miliband R, Panitch L (eds) Real problems, false solutions: socialist register 1993. Merlin Press, London, pp 1–51

Heynen N (2014) Urban political ecology I: the urban century. Prog Hum Geogr 38(4):598–604
Heynen N, Kaika M, Swyngedouw E (eds) (2006) In the nature of cities: urban political ecology and the politics of urban metabolism. Routledge, London
Holling CS (1973) Resilience and stability of ecological systems. Annu Rev Ecol Syst 4:1–23
Hornborg A (2009) Zero-sum world: challenges in conceptualizing environmental load displacement and ecologically unequal exchange in the world-system. Int J Comp Sociol 50(3–4):237–262
ICLEI (2018) The ICLEI Montréal action plan 2018–2021. ICLEI, Bonn
Ingalls ML, Stedman RC (2016) The power problematic: exploring the uncertain terrains of political ecology and the resilience framework. Ecol Soc 21(1):6
Kaika M (2017a) 'Don't call me resilient again!': The new urban agenda as immunology ... or ... what happens when communities refuse to be vaccinated with 'smart cities' and indicators. Environ Urban 29(1):89–102
Kaika M (2017b) Between compassion and racism: how the biopolitics of neoliberal welfare turns citizens into affective 'idiots'. Eur Plan Stud 25(8):275–1291
Kaika M, Swyngedouw E (2014) Urban political ecology – Great promises, deadlock ... and new beginnings? Doc Anal Geogr, Barcelona, UAB 60(3):459–481
Keil R (2003) Urban political ecology 1. Urban Geogr 24(8):723–738
Keil R (2005) Progress report – urban political ecology. Urban Geogr 26(7):640–657
Leichenko R (2011) Climate change and urban resilience. Curr Opin Environ Sustain 3(3):164–168
Loftus A (2012) Everyday environmentalism: creating an urban political ecology. University of Minnesota Press, Minneapolis
MacKinnon D, Derickson KD (2012) From resilience to resourcefulness: a critique of resilience policy and activism. Prog Hum Geogr 37(2):253–270
Magis K (2010) Community resilience: an indicator of social sustainability. Soc Nat Resour 23-5:401–416
March H (2018) The smart city and other ICT-led techno-imaginaries: any room for dialogue with degrowth? J Clean Prod 197:1694–1703
March H (2022, forthcoming) Cities and the environment. In: Pellizzoni L, Leonardi E, Asara V (eds) Handbook of critical environmental politics. Edward Elgar Publishing, Cheltenham
Marí-Dell'Olmo M, Tobías A, Gómez-Gutiérrez A, Rodríguez-Sanz M, García de Olalla P, Camprubí E, Gasparrini A, Borrell C (2018) Social inequalities in the association between temperature and mortality in a South European context. Int J Public Health 64(1):1–11
Matin N, Forrester J, Ensor J (2018) What is equitable resilience? World Dev 109:197–205
Matyas D, Pelling M (2015) Positioning resilience for 2015: the role of resistance, incremental adjustment and transformation in disaster risk management policy. Disasters 39(SI):1–18
Meerow S, Neuner FG (2021) Positively resilient? How framing local action affects public opinion. Urban Aff Rev 57(1):70–103
Meerow S, Newell JP, Stults M (2016) Defining urban resilience: a review. Landsc Urban Plan 147:38–49
Meerow S, Pajouhesh P, Miller TR (2019) Social equity in urban resilience planning. Local Environ 24(9):793–808
Mees HLP, Uittenbroek CJ, Hegger DLT, Driessen PPJ (2019) From citizen participation to government participation: an exploration of the roles of local governments in community initiatives for climate change adaptation in the Netherlands. Environ Policy Gov 29(3):198–208
Muñoz-Erickson TA, Miller CA, Miller TR (2017) How cities think: knowledge co-production for urban sustainability and resilience. Forests 8:203
Nadasdy P (2007) Adaptive co-management and the gospel of resilience. In: Armitage D, Berkes F, Doubleday N (eds) Adaptive co-management: collaboration, learning and multi-level governance. UBC Press, Vancouver, pp 208–227
Nelson DR (2011) Adaptation and resilience: responding to a changing climate. WIREs Clim Change 2(1):113–120

Nelson SH (2015) Beyond the limits to growth: ecology and the neoliberal counterrevolution. Antipode 47(2):461–480
Norris FH, Stevens SP, Pfefferbaum B, Wyche KF, Pfefferbaum RL (2008) Community resilience as a metaphor, theory, set of capacities, and strategy for disaster readiness. Am J Community Psychol 41:127–150
Otu A, Ahinkorah BO, Ameyaw EK, Seidu AZ, Yaya S (2020) One country, two crises: what Covid-19 reveals about health inequalities among BAME communities in the United Kingdom and the sustainability of its health system? Int J Equity Health 19:189
Peck J, Tickell A (2002) Neoliberalizing space. Antipode 34(3):380–404
Pellizzoni L (2015) Ontological politics in a disposable world: the new mastery of nature. Ashgate, Farnham
Peterson G (2000) Political ecology and ecological resilience: an integration of human and ecological dynamics. Ecol Econ 35(3):323–336
Pike A, Dawley S, Tomaney J (2010) Resilience adaptation and adaptability. Camb J Reg Econ Soc 3:59–70
Porter L, Davoudi S (2012) The politics of resilience for planning: a cautionary note. Plan Theory Pract 13(2):329–333
Protevi J (2013) Life, war, earth: Deleuze and the sciences. University of Minnesota Press, Minneapolis
Robbins P (2012) Political ecology: a critical introduction, 2nd edn. Wiley-Blackwell, New York
Ruiz-Mallén I, Corbera E (2013) Community-based conservation and traditional ecological knowledge: implications for socio-ecological resilience. Ecol Soc 18(4):12
Sapountzaki K (2007) Social resilience to environmental risks: a mechanism of vulnerability transfer? Manag Environ Qual Int J 18:274–297
Satorras M, Ruiz-Mallén I, Monterde A, March H (2020) Co-production of urban climate planning: insights from the Barcelona climate plan. Cities 106:102887
Saurí D (2018) From sustainability to resilience: the hidden costs of recent socioenvironmental change in cities of the global north. In: Ward K, Jonas AEG, Miller B, Wilson D (eds) The Routledge handbook on spaces of urban politics. Routledge, New York
Stevenson F, Petrescu D (2016) Co-producing neighbourhood resilience. Build Res Inf 44(7):695–702
Stumpp EM (2013) New in town? On resilience and "resilient cities". Cities 32:164–166
Suárez M, Gómez-Baggethun E, Benayas J, Tilbury D (2016) Towards an urban resilience index: a case study in 50 Spanish cities. Sustain 8(8):774
Swyngedouw E (2006) Circulations and metabolisms: (hybrid) natures and (cyborg) cities. Sci Cult 15(2):105–121
Swyngedouw E (2022a) Capital's natures: a critique of (urban) political ecology. In: Tzaninis Y, Mandler T, Kaika M, Keil R (eds) Pumping up the heat: urban political ecology. The University of Manchester Press (forthcoming), Manchester
Swyngedouw E (2022b) Illiberalism and the democratic paradox: the infernal dialectic of neoliberal emancipation. Eur J Soc Theory 25(1):53–57
Swyngedouw E, Ernstson H (2018) Interrupting the Anthropo-obScene: immuno-biopolitics and depoliticising ontologies in the Anthropocene. Theory Cult Soc 35(6):3–30
Swyngedouw E, Heynen N (2003) Urban political ecology, justice and the politics of scale. Antipode 34(4):898–918
Turner MD (2014) Political ecology I: an alliance with resilience? Prog Hum Geogr 38(4):616–623
Vale LJ (2014) The politics of resilient cities: whose resilience and whose city? Build Res Inf 42(2):37–41

Walker J, Cooper M (2010) Genealogies of resilience: from systems ecology to the political economy of crisis adaptation. Secur Dialogue 42(2):143–160

Walker B, Holling CS, Carpenter SR, Kinzig A (2004) Resilience, adaptability and transformability in social–ecological systems. Ecol Soc 9(2):5

Wilbanks TJ, Kates RW (2010) Beyond adapting to climate change: embedding adaptation in responses to multiple threats and stresses. Ann Assoc Am Geogr 100(4):719–728

Chapter 2
Bridging Urban Climate Justice and Participatory Governance to Explore the Transformative Capacity of Climate Resilience

Kaitlin Strange, Mar Satorras, and Hug March

Abstract With increasing threats due to climate change, scholars and practitioners turn to cities and their stakeholders as key players in climate governance and action. At the same time, there is an emerging body of literature that looks at the transformative potential of inclusive urban climate adaptation. This chapter sets out to examine and situate urban transformative adaptation by exploring its connections and differences with two bodies of literature: (1) urban climate justice; and (2) community-driven and participatory urban climate governance. We first review climate governance and the importance of cities when preparing for climate change. Additionally, we follow the origins and evolution of climate justice and the call for transformative and equitable climate adaptation in cities. This review of transformative adaptation literature highlights the importance of participatory and inclusive processes. Therefore, the chapter looks at how civil society and community-based organisations engage in urban climate adaptation. We argue that a bottom-up and community-lead approach to resilience and adaptation planning can address local needs, ensure equity and justice, and challenge the status quo in order for transformative change to take place.

K. Strange (✉)
Internet Interdisciplinary Institute (IN3), Universitat Oberta de Catalunya (UOC), Barcelona, Spain
e-mail: kstrange@uoc.edu

M. Satorras
Internet Interdisciplinary Institute (IN3), Universitat Oberta de Catalunya (UOC), Barcelona, Spain

Barcelona Institute of Regional and Metropolitan Studies (IERMB), Universitat Autònoma de Barcelona, Bellaterra, Spain

H. March
Estudis d'Economia i Empresa and Internet Interdisciplinary Institute (IN3), Universitat Oberta de Catalunya (UOC), Barcelona, Spain

I. Ruiz-Mallén et al. (eds.), *Urban Resilience to the Climate Emergency*, The Urban Book Series, https://doi.org/10.1007/978-3-031-07301-4_2

Keywords Climate justice · Climate resilience · Community-based adaptation · Transformative adaptation · Urban climate governance

2.1 Introduction

The past century has seen intense urbanisation with 68% of the world's population anticipated to live in cities by 2050 (UN 2019). Our cities continue to experience threats from climate change, such as flooding, sea level rise, drought, and extreme heat events (Rosenzweig et al. 2015). And given the prospects laid out in the last report of the Intergovernmental Panel on Climate Change (IPCC) those threats will be exacerbated in the coming decades. However, climate change does not affect everyone equally (IPCC 2021). The poor and disenfranchised suffer negative impacts of climate change disproportionately while receiving fewer benefits from adaptation and resilience efforts (Anguelovski et al. 2016; Hoffman et al. 2020). Growing awareness of the uneven impacts of climatic threats leads scholars and practitioners to critically examine who benefits and who is negatively impacted. Considering the interconnected and complex risks cities face, there is a call for radical, long-term, and transformational strategies that addresses the underlying systemic inequalities and conditions for climate vulnerability (IPCC 2012; Chu et al. 2019; Hughes and Hoffmann 2020; Wilson et al. 2020; Shi and Moser 2021). As the world faces the challenges of urbanisation, changing climates, and growing social and environmental inequities, there is a growing debate around adapting our cities in ways that are just, sustainable, and transformative.

This chapter reviews the main concepts and literature that have paved the way for urban transformative climate adaptation and resilience. This review draws on literature from climate justice, climate governance, and urban studies, to highlight the characteristics, potential, and limitations of urban transformative adaptation. As a complex and interconnected concept/approach, urban transformative climate adaptation requires bridging different fields and theories. To examine and situate the approach of urban transformative adaptation, this chapter explores its connections and differences with two bodies of literature: (1) urban climate justice; and (2) community-driven and participatory urban climate governance. In particular, we illuminate and trace the evolution of urban climate governance, discussing the importance of inclusion and participation in equitable climate planning, and engaging with the growing fields of climate justice and transformative adaptation at the urban level. In doing so, we examine the role of community groups in the adaptation process and explore their potential as agents of transformative change.

The following section presents an overview of urban climate governance and the increasing role cities play in developing climate policy and planning for adaptation and resilience. The third section of this chapter explores theories and research around equitable climate adaptation and climate justice, bringing to light a growing body of work around transformative urban adaptation. The fourth section examines the role of civil society and the transformative potential of participatory, co-produced and community-based adaptation. The chapter ends by highlighting the characteristics, potential, and limitations of transformative climate adaptation and discussing the potential community-driven and bottom-up initiatives for adapting to climate change in an equitable and just manner.

This chapter provides a theoretical and conceptual basis to better understand the transformative potential and related barriers for radical transformation as is identified in the case studies on the institutional efforts for urban climate resilience and the bottom-up and co-produced initiatives included in Parts II and III of this book, respectively. Furthermore, this chapter reviews and provides insights for transformative approaches to urban climate adaptation and resilience by exploring the intersections of climate justice theory and participatory and community-led adaptation research.

2.2 Tackling Climate Change Through Urban Climate Adaptation and Governance

2.2.1 Adaptation and Urban Climate Governance

The field of climate adaptation governance has grown since its inception. The need to focus on adaptation has progressively become clear as the world continues to experience more dramatic impacts of climate change (Schipper 2006; Aylett 2015). In 2021, the IPCC published the first deliverable by Working Group 1 (*The Physical Science basis*) of the Sixth Assessment Report which points to a daunting future of extreme climate events and changing landscapes (IPCC 2021). This last report states that global surface temperature will continue to increase, for at least the next 30 years, whether countries act against carbon dioxide (CO2) emission or not. Furthermore, the report emphasises that global warming of 1.5 °C and 2 °C will be exceeded, unless significant reductions in CO2 and other greenhouse gas emissions occur soon (IPCC 2021). The report outlines an expected increase in extreme climate events such as heatwaves, heavy precipitation, droughts, and tropical cyclones (IPCC 2021). This new evidence reinforces the need to plan for climate adaptation, which previous reports have already identified as a key priority for urban governments (Bazaz et al. 2018).

The release of the first deliverable of the IPCC Sixth Assessment Report coincided with the UN Climate Change Conference in 2021 (COP26). The urgency brought about from the report can be seen in the COP26's four goals: (1) secure global net zero by mid-century and keep 1.5° within reach; (2) adapt to protect communities and natural habitats; (3) mobilise finance to fund carbon neutrality and climate adaptation; and (4) work together to address the challenges of the climate crisis (COP26 2021). At COP26, subnational authorities, including cities, towns, and regions, had a home in the Multilevel Action Pavilion. Local governments went to Glasgow with a four-point roadmap, which included: (1) multilevel action as the new normal for the next phase of the Paris Agreement; (2) localising national, global, and private finance to seize urban opportunities for bold action in cities and regions; (3) just climate action for all; and (4) stepping up subnational engagement in the UNFCCC processes (LGMA 2021). However, after COP26, local governments networks considered that the resulting Glasgow Climate Pact failed *to capture the emergency mode that the moment called for* and to provide *a clear vision for multilevel and cooperative action*, while envisioned next COP27 as crucial to address the role of cities in climate adaptation (Arikan 2021).

With adaptation at the forefront of the debate around climate change, scholars have turned to examine how adaptation is planned for and implemented in cities (Hughes 2020; Meerow and Woodruff 2020); who are the key stakeholders and what are the power dynamics at play (Klein et al. 2018; Woroniecki et al. 2019; Kauffman and Hill, 2021); and how is equity, inclusion, and justice addressed in these urban adaptation efforts (Mallory and Ashcroft 2020; Coggins et al. 2021; Fiack et al. 2021; Mohtat and Khirfan 2021).

Cities have often been neglected in climate discussions, where the debate has been centred on a global perspective (Castán Broto and Bulkeley 2013; Fuhr et al. 2018). Nonetheless, the international and national policy debate is shifting towards an urban focus, as seen in the increasing prioritisation and articulation of urban-related goals of the IPCC, the United Nations (UN) Sustainable Development Goals (SDGs), the UN Office for Disaster Relief's Sendai Framework, and other policy and research institutions (IPCC 2014; Rosenzweig et al. 2015; UNISDR 2015; United Nations General Assembly 2015; World Climate Research Programme 2019). Thus, cities have become a focal point in the debate about climate change and are increasingly seen as key players in planning for a future of environmental uncertainty (Rosenzweig et al. 2015; Romero-Lankao et al. 2018; Long and Rice 2019; Bulkeley 2021). Municipal governments play a significant role in tackling pressing challenges by informing local, national, and international policy and are the focus of a growing number of climate planning initiatives (Whitehead 2013; IPCC 2014). Municipalities across the world are forming alliances and networks, such as the UN's Conference of the Parties, Global Covenant of Mayors for Climate and Energy, 100 Resilient Cities, and C40 Cities to address climate change by sharing best practices, disseminating information and research, and collaborating on policy and planning (Rosenzweig et al. 2015; 100 Resilient Cities 2020; C40 2020; Global Covenant of Mayors 2020).

With the growing role of cities in mitigating and preparing for climate change, urban climate governance scholarship has been expanded (Castán Broto and Westman 2020). Urban climate governance refers to how public, private, and civil actors and institutions enact, influence, and manage urban climate goals, planning and implementation (Anguelovski and Carmin 2011). There is a rich body of work that seeks to understand the development and performance of urban climate governance in cities (der Heijden 2019). This literature focuses on the potential of experimentation, innovation, and multi-level networks in addressing climate change and its threat to cities (Bulkeley and Castán Broto 2013; Busch et al. 2018; Gordon 2018). Successful and effective urban climate governance needs to be accountable and transparent, responsive, and flexible, participatory and inclusive, and supported by leadership with demonstrated experience (Sarzynski 2015). Urban climate governance debates seek to tackle climate change through different approaches and scales including: mitigation and adaptation; local and regional; policy and implementation; research and practice (Anguelovski and Carmin 2011; der Heijden 2019; Castan-Broto and Westman 2020).

Scholars and policy makers are also increasingly using a climate urbanism framework, which refers to *how urban areas are lived, and how they are governed and imagined in the climate emergency* (Castán Broto and Robin 2021, p. 2). Critical scholarship has examined how climate urbanism is framed and practised in contemporary cities, to develop a critical view of market-based approaches and the possibility of such urbanism to furthering inequalities (Long and Rice 2019; Castán Broto and Robin 2021). With a focus on adaptation, political ecologists and critical urban scholars have contributed to disentangle the complex relation between social inequality, urbanisation and uneven exposure to climate (and broadly speaking socio-environmental) risks (Cook and Swyngedouw 2012; Angelo and Wachsmuth 2015; Anguelovski et al. 2016; Reckien et al. 2017). Urban vulnerability to climatic hazards is complex and intertwined with urban infrastructure, economy, politics, ecology, health, and culture (Chu et al. 2019). Changing socio-demographic structures and urban configurations combined with the impending effects of the climate emergency require cities to seek out holistic approaches and innovative solutions to address both physical climate threats as well as the potential social impacts of a changing environment (UN-Habitat 2011; Sarzynski 2015; Hölscher et al. 2019b; Elmqvist et al. 2019; Castán Broto and Westman 2020; McPhearson 2020; Sengers et al. 2021). However, there is no one-size fits all approach to urban climate governance for adaptation and climate resilience. Actions are planned and implemented based on local context, priorities, and government structures (Castán Broto 2017). A review of past and recent literature on urban climate resilience and adaptation provides insight into the characteristics, nuances, and foundational theory that underpins this debate.

2.2.2 Urban Climate Adaptation and Resilience Planning

City governments look to adaptation planning to limit urban vulnerability to climate change, align with existing international agendas, or to enhance the city's image (Anguelovski and Carmin 2011). Urban climate adaptation planning includes five stages: identification and vulnerability assessment; planning; implementation of strategies; monitoring and evaluation; and revision and sharing (Nordgren et al. 2016). On the ground, urban climate adaptation is implemented through hard or grey infrastructure (e.g., seawalls, levees, storm sewers), green infrastructure (e.g., urban parks, bioswales, urban forests, constructed wetlands, green roofs), and soft infrastructure (e.g., knowledge, information, educations, incentives) (IPCC 2012; Moser and Pike 2015).

Key actors working on climate adaptation at the urban scale include municipal staff and elected officials, residents, community organisations (e.g., NGOs, research organisations, informal groups, grassroots movements), private corporations, and regional, national, and global institutions (planning bodies, coalitions, financiers, national governments and ministries, international NGOs, city networks) (Zografos et al. 2020). These stakeholders engage in decision-making processes, both formal and informal, that work to prepare and adapt cities in the face of climate uncertainty (Boyd and Folke 2011).

Urban climate adaptation is often framed within the context of urban resilience planning and theory, which is grounded in the study of complex socio-ecological systems. Urban resilience is applied to different risks ranging from the economy, communities, infrastructure, public health, to climate change (Meerow et al. 2016). Resilience theory, founded in the ecology, psychology, engineering fields, attempts to understand how complex systems continue to function in the face of change and stress (Meerow et al. 2019). The debate has evolved into the urban setting to address functionality and resilience in the face of natural disasters, international development, economic downturn, and climate change (Woodruff et al. 2018). In what regards resilience to climate change impacts, there are four typical framings: urban shock-proofing (short-term & system focus), resilience planning (long-term & system focus), community disaster resilience (short-term & community focus), and resilient community development (long-term & community-focus) (Wardekker 2021).

Scholarship and practice of urban resilience and adaptation are, at times, criticised for their limited discussion of equity and social vulnerability (Pelling 2011; Boyd and Juhola 2015; Ziervogel et al. 2017; Meerow et al. 2019). Researchers are turning a critical lens to widely praised urban climate resilience plans and policies, calling out their uneven benefits and unintended, albeit real, negative outcomes. Climate resilience plans in cities such as Philadelphia, Boston, Malmö or Barcelona, among others, have been criticised for paving the way for the development of green infrastructure in a way that may cause residential and social displacement, or in other words *climate gentrification* (Anguelovski et al. 2019). Despite their popularity, critics question the long-term social value of urban climate resilience and adaptations plans and interventions (Anguelovski et al. 2016; Kaika 2017; Matin et al. 2018).

In this chapter, while mobilising a critical perspective on urban climate resilience, we focus on equitable climate adaptation's transformative potential in urban areas. A critical look at climate resilience and urban adaptation lays the groundwork for further research on equitable resilience planning and climate adaptation. The following section highlights the influence of environmental justice on the current debate around climate equity and the potential of transformative adaptation in creating a more just and sustainable future in the face of climate change.

2.3 Climate Justice and Transformative Adaptation: A Call for Equitable and Just Urban Climate Governance

The discussion around urban climate adaptation and resilience continues to evolve, while articulating and mobilising different disciplines and theories. Literature at the intersection of climate change and social equity brings together frameworks, values, and critiques from justice scholarship and advocacy, equitable planning practices, and just transition theory. The result is a growing discussion around climate justice and the potential for transformative change in the form of urban adaptation planning and implementation. The following section examines the influence of environmental justice on the climate justice debate in the urban context, the contributions of the growing field of urban transformative adaptation, and untangles the differences and similarities between the two.

2.3.1 Urban Environmental and Climate Justice

As mentioned above, climate change negatively impacts vulnerable populations disproportionately. The environmental justice movement has struggled against these urban environmental injustices since the early 1980s when communities rallied against hazardous waste in low-income neighbourhoods (Bryant and Mohai 1992). The initial debate focused on inequitable distribution of environmental *bads*, where some communities receive more environmental risks than others. The field grew to address the uneven distribution of environmental *goods* as well, such as green infrastructure, urban parks, and sustainable development (Schlosberg 2013) and nowadays covers an array of issues, including distribution of transit, food, water, jobs, brownfields, and climate change impacts (Cook and Swyngedouw 2012).

Climate justice evolved out of the environmental justice movement and addresses the uneven impact of current and expected risks of climate change (Schlosberg 2012). Climate vulnerabilities and inequalities are exacerbated by the city's dense networks of interwoven socio-spatial processes. Vulnerability of the urban poor is based not only on exposure to climate-related risks and hazards, but also on systematic and structural conditions such as economic disparity, lack of representation, limited opportunities, and a lack of services (Pelling 2011; Schlosberg 2012; Manuel-Navarrete and Pelling 2015; Castán Broto 2017).

Urban climate justice scholarship sheds light on the ways that urban populations are impacted differently by climate change depending on social, economic, and individual factors (Reckien et al. 2017). To understand the complexity of these socio-environmental realities, urban scholars examine physical, economic, political, and social processes that shape urbanisation (Swyngedouw and Heynen 2003). Along these lines, urbanisation is seen as an expression of broader capitalist processes shaped by social and political power (Whitehead 2013). In other words, the processes behind socio-environmental urban conditions create different outcomes for different individuals and communities (Eriksen et al. 2015; Hoffman et al. 2020), and this impacts how climate change unevenly affects a given community and its adaptive capacity.

The climate justice debate began by examining distributive justice, specifically mitigation requirements (Shue 1993; Coggins et al. 2021; Fiack et al. 2021), historical responsibility and restorative justice (Agarwal and Nurain 1991), and per capita emission allowances (Jamieson 2001; Singer 2004). Scholarship on climate justice has since evolved to include analyses of the negative impacts of climate change or the unequal distribution of adaptation initiatives, climate reparations, and sacrifice zones (Schlosberg 2013; Anguelovski et al. 2019; Zografos and Robbins 2020; Sultana 2021).

In cities, low-income residents, people of colour, and migrant communities are often hit hardest by the uneven distribution of climate risks and have less access to adaptation benefits. These populations contribute the least to climate change, have the least access to environmental amenities, are the most exposed to climate hazards, lack resources to adapt, and have been victim to *climate gentrification* (Anguelovski et al. 2019). At times, urban climate adaptation planning falls short of long-term socially just outcomes by both favouring privileged groups and denying resources or a voice to marginalised communities through provisions of protective infrastructure, enforcement of land use regulations, inclusion in the planning process, and engagement with the private sector (Anguelovski et al. 2016). By employing a perspective rooted in climate justice and participation, research can push for a more human-centred agenda to further urban goals of transformative urban change and a more equitable future (Anguelovski et al. 2019; Mikulewicz 2019). In what follows, we review the conversation around urban transformative adaptation examining its goals, limitations, potential, and its relationship to climate justice.

2.3.2 Urban Transformative Adaptation

The field of urban transformative adaptation, underpinned by climate justice values, is growing and evolving in response to critical examinations of climate adaptation. This area of study exposes conventional climate governance's role in reinforcing social inequalities and sets out to address the need for a more intersectional, relational, and emancipatory approach (Anguelovski et al. 2020; Revi et al. 2020). By applying a transformative lens to climate adaptation, it is possible to further integrate the debates around social justice and climate change and analyse the roots of climate change in economic and social injustice (Newell et al. 2021).

As mentioned in previous section, the current adaptation and climate resilience debate is criticised for its lack of attention to the root causes of climate vulnerability and the interconnectedness of climate change, uneven development, and power dynamics (Schlosberg et al. 2017; Romero-Lankao et al. 2018, Shi and Moser 2021). Many approaches to urban climate adaptation focus on incremental solutions that mitigate near-term risks, employing minor or small-scale strategies and interventions which support business-as-usual (Kates and Wilbanks 2012; Wolfram et al. 2016; Romero-Lankao et al. 2018; Chu et al. 2019; Fedele et al. 2019; Hölscher et al. 2019a). Incremental adaptation, many argue, does not address underlying systemic issues, can increase costs, and delay the implementation of long-term, sustainable, and equitable solutions (Fedele et al. 2019; Wilson et al. 2020). Furthermore, urban adaptation processes and projects can further exacerbate inequalities and displacement of vulnerable groups and residents. Examples of adaptation planning's negative consequences can be seen in housing displacement from urban greening and privatisation of utility services and subsequent price increases (Revi et al. 2020).

In response to the limitations and insufficiency of top-down, techno-managerial, and incremental approaches to adaptation, transformative adaptation scholarship and practice continues to grow (Ribot 2014; Nightingale 2017). Transformative adaptation gained visibility with IPCC's special report on Managing Risk of Extreme Events and Disasters to Advance Climate Change Adaptation (IPCC 2012), which called for *transformation* to manage risk and adapt to climate extremes (Pelling et al. 2015). Transformative adaptation, also referred to as *radical adaptation* (Dawson 2017), seeks to restructure political economies at multiple scales and calls for a significant shift in power relations (Pelling 2011; Kates and Wilbanks 2012; Bassett and Fogelman 2013; Zografos et al. 2020). When transformative adaptation is approached by *just transition* theory, scholars examine how processes of change, alternative futures, and political or structural dynamics in the name of justice informed transitions and transformation (Hughes and Hoffmann 2020).

Transformative adaptation is often discussed in vague terms with broad implications and aspirations, also when is specifically applied in urban areas (Elmqvist et al. 2019). This is changing as scholars increasingly strive to characterise and operationalise transformative adaptation. For instance, Fedele et al. (2019, p. 116) define transformative adaptation as the *fundamental systems' changes that address root causes of vulnerability* to climate change. In the context of cities, Chu et al. (2019) state that transformative adaptation *focuses on systemic changes to development processes that improve people's quality of life; enhance the social and economic vibrancy of cities; and ensure sustainable, resilient, and inclusive urban futures* (Chu et al. 2019, p. 12). To pin down a working definition the literature breaks down the characteristics and attributes of transformative adaptation, as well as the conditions needed for transformative adaptation to take place. For instance, Fedele et al. (2019) and Deubelli and Mechler (2021) consider that transformative adaptation is characterised as restructuring, path-shifting, innovative, multiscale, systemwide, and persistent.

In the urban context, transformative adaptation gained prominence after the Fifth Assessment Report of the IPCC which pointed at the potential of urban adaptation to address both the root causes of poverty and the failures in sustainable development (Revi et al. 2014). A handful of cities, including Rotterdam, Barcelona, New York City, or San Francisco have released climate action and adaptation projects infused by the essence of transformative adaptation, seeking to provide co-benefits through climate adaptation, greening, recreation, community-building and economic development (Hölscher et al. 2019a; Satorras et al. 2020; Hamel et al. 2021).

While there is growing interest and research dedicated to urban transformative adaptation, a critical examination of the field is emerging. Acknowledging the potential behind systemic and radical change, scholars also remind us that transformation is inherently political and requires academics and practitioners alike to ask: transformation of what, for whom, and for what purpose (Blythe et al. 2018; Ajibade and Adams 2019; Newell et al. 2021). If power dynamics and political interests are not addressed, there is a risk of reproducing and reinforcing unequal dynamics under the veil of transformation (Woroniecki et al. 2019; Kuhl et al. 2021). Furthermore, the potential of transformative adaptation can be challenged by unclear distribution of responsibilities, lack of integration, high levels of investment, or the prioritisation of short-term economic interests (Pelling 2011; Nordgren et al. 2016; Chu et al. 2017; Romero-Lankao et al. 2018; Zografos et al. 2020).

Deep-rooted changes that challenge the status quo and existing systems are needed for large scale transformation to take place. Strong leadership and authentic stakeholder engagement must be present for this change to occur (Deubelli and Mechler 2021). This understanding of adaptation is founded on a desire to reform and remove the structures that drive vulnerabilities and the systems that place an uneven burden of climate risk and response (Newell et al. 2021). Transformative adaptation pathways present a model of decision making based on strong leadership; inclusion and equity; finance and local capacity; synergies across regional, national, and global scales; knowledge, data, and partnerships; evaluation and learning; accountable institutions and governance (Chu et al. 2019; Iwaniec et al. 2019). In the vein of breaking down silos and approaching adaptation from an intersectional and holistic manner, scholars are calling for climate transformations to better integrate mitigation and adaptation planning and policies to create a future that is both socially just and environmentally sustainable (Hurliman et al. 2021).

An overview of both transformative adaptation and climate justice presents a critical guide to future equitable climate planning and action. There is value in understanding the foundational theories behind these similar but divergent approaches. The section that follows outlines the similarities and differences between climate justice and transformative adaptation, providing further insight into the potential of this growing literature in the urban context.

2.3.3 The Intersection of Climate Justice and Urban Transformative Adaptation

Climate justice is grounded in a long tradition of scholarship and practice that strives to ensure a more equitable future for all. As mentioned above, climate justice scholarship is largely based on an examination of distributive, recognition, and procedural justice (Holland 2017; Fainstein 2018; Svarstad and Benjaminsen 2020). This approach helps illustrate the ways in which vulnerable and marginalised communities are negatively impacted by both climate change's threats and solutions. In parallel, there is growing literature that explores a more systemic and transformative approach to climate justice and equitable adaptation. In this regard in Fig. 2.1 we synthetise the interconnections and continuities between climate justice and urban transformative adaptation scholarship.

There is a clear connection between the fields of climate justice and transformative adaptation, in that they emphasise the need for people-centric planning, processes that are historically and culturally responsive, and a need to address inequities and injustices in the allocation and distribution of services, resources, and infrastructure. Urban transformative adaptation literature incorporates core concepts of climate justice and overlays a framework that focuses on coordination and partnerships, innovation and experimentation, and financial systems and incentives (Romero-Lankao et al. 2018; Chu et al. 2019; Zografos et al. 2020; Shi and Moser 2021). As Fig. 2.1 outlines, transformative adaptation builds off climate justice frameworks by broadening the scale, timeframe, and scope. Such approach often looks at the processes of adaptation planning and decision making as adaptation pathways. By integrating the metaphor of *pathways* within adaptation frameworks, scholars and planners are able to visualise processes, sequences, or series of adaptation decisions over time (e.g., Haasnoot et al. 2013; Maru et al. 2014; Wise et al. 2014). This growing body of literature brings together a larger perspective, connecting many fields and practices to assess how policies, plans, and projects are affected and enacted at different scales and across different domains (Newell et al. 2021). In summary, transformative adaptation sets out to expand and operationalise the philosophies and theories behind climate justice.

Transformative adaptation and equitable planning rely not only on flexible, long-term, and innovative climate governance (Chu et al. 2019; Deubelli and Mechler

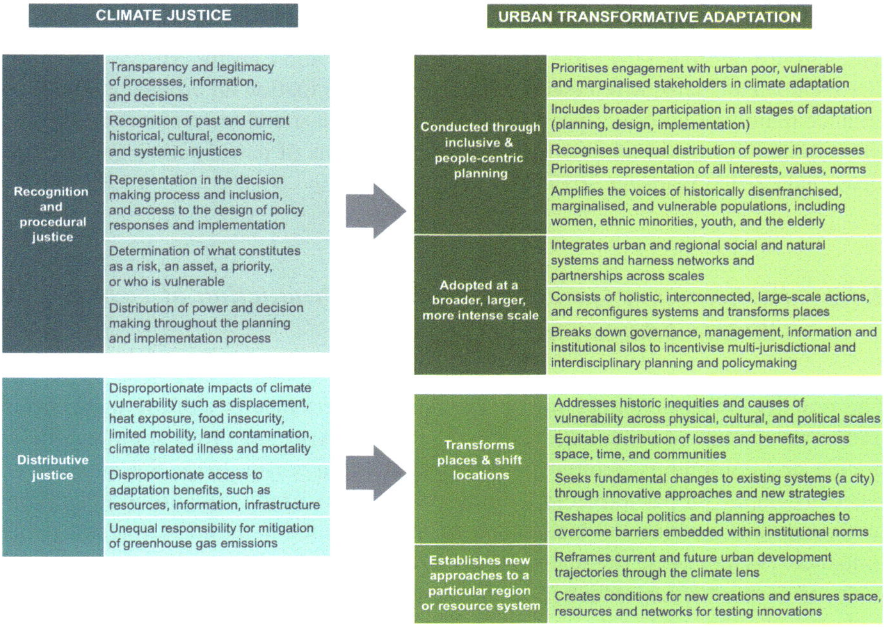

Fig. 2.1 Key characteristics of and connections between climate justice and urban transformative adaptation. (Source: own elaboration, based on Kates and Wilbanks (2012), Bulkeley et al. (2013), Reckien et al. (2017), Romero-Lankao et al. (2018), Chu et al. (2019), Hölscher et al. (2019a, b), Zografos et al. (2020), Shi and Moser (2021))

2021). There is growing scholarship that emphasises the need for participatory, community-lead, and decentralised approaches to adaptation to create the conditions necessary for transformation. By examining civil society's involvement in climate governance and equitable climate adaptation, we can begin to identify the potential and limitations of community-led efforts and co-production in urban climate planning for adaptation and resilience.

Scholarship around climate justice and transformative adaptation centres community participation as an essential pillar of equitable and sustainable change that goes beyond business as usual. The next section outlines the ways communities engage in urban adaptation and climate resilience planning by examining the tactics, incentives, and types of community-led initiatives. In defining and characterising the role of community and non-governmental organisations we start to see their potential to envision, inform, and lead transformative urban adaptation.

2.4 The Role of Civil Society in Urban Climate Adaptation and Resilience

A socially focused approach to urban adaptation planning relies on an examination of the role of communities and organisations. Scholars see participation as having intrinsic value as well as being a tool to ensure a more representative, transformative, and just approach to climate risks (Anguelovski and Carmin 2011; Schlosberg 2013; Chu et al. 2019; Mikulewicz 2019). Civic participation on urban climate governance manifests across a spectrum moving between social movements and activism, community-based adaptation, and different forms of participatory governance (Sarzynski 2015; Wamsler 2017; Mees et al. 2019).

2.4.1 Participatory Governance and Co-Production of Urban Climate Adaptation

The practice of participatory urban adaptation governance is shifting away from traditional models of consultation and information (Arnstein 1969) to more collaborative and community-centred planning tactics (Sarzynski 2015; Satorras 2021; see also Chap. 7 for participatory wildfire governance). Traditional urban adaptation participation is government-lead and is generally limited in duration, intensity, and influence. Further along the scale of citizen participation we can find inclusive planning and co-production, which is often connected to a partnership model that comes with expectations for high impact from citizens on decision-making (Wamsler 2017; Ruiz-Mallén 2020; Satorras et al. 2020). Although top-down and bottom-up approaches are both widely recognised, the actions in practice are often combinations of these approaches (IPCC 2014; see also Chap. 6 for nature-based solutions in schoolyards stemming from collaboration between the educational community and municipal government authorities).

Government engagement in adaptation initiatives may also vary from regulating and steering roles towards more collaborative approaches enabling and facilitating community initiatives that can be either co-produced with or self-governed by citizens (Mees et al. 2019). The next sections deep into activism and collective action, community-based adaptation, and do-it-yourself initiatives, in which local governments may also engage in such different roles.

2.4.2 *Social Movements and Collective Action*

Social and environmental movements alongside grassroots organisations have worked for decades to create awareness around environmental issues and advocate for sustainable, ethical, and ecological change (Mix 2011). Since the 1990s, the climate movement has become a central rallying point for environmentalists and human rights activists alike to speak out against climatic risks and its ecological and sociological impacts, as seen during mobilisations around the 2009 United Nations Climate Change Conference, the 2014 People's Climate March, and the more recent Global Climate Strike (Jamison 2010; Castán Broto and Westman 2020; de Moor et al. 2021). Youth climate activists have organised marches, protests, and sit-ins alongside events and movements such as Extinction Rebellion, Sunrise Movement, and the global #FridaysforFuture network (Fisher 2019; Schinko 2020). These youth-lead events build on the broader climate movements while engaging a new population of young activists. There is growing research examining the role of social movement and collective action in furthering climate action, and particularly sustainable and socially just action (Caniglia et al. 2015; de Moor et al. 2021; see also Chap. 8 for the role of new climate movements in defining the climate emergency action plan in Barcelona). Social movements and grassroots activism illustrate the power of organisations to push an agenda of radical change and be agents of change in the face of climate threats (Pellow 2001; see also Chap. 9 for the analysis of self-organised civil society initiatives contributing to urban climate resilience in Seville).

2.4.3 *Community-Based and Do-It-Yourself Climate Adaptation*

There is a growing body of research in the field of participatory governance that sheds light on environmental and social organisations' role in community-based and do-it-yourself (DIY) adaptation. Based in international and development studies, community-based adaptation (CBA) is an approach to adaptation that acknowledges the skills, experience, local knowledge, and networks of local communities and relies on community groups to take on locally appropriate activities (Anguelovski and Carmin 2011; Dodman and Mitlin 2013). CBA, if supported by city agencies and mainstreamed, can lead to effective risk management, increased local capacity, and enhanced government accountability (Archer et al. 2014). However, like other forms of participatory governance, CBA efforts often lack funding, buy-in, authentic participation, and a lack of clear and agreed priorities (Few et al. 2007; Dodman and Mitlin 2013; Forsyth 2013; Archer et al. 2014).

Citizens also find ways to take urban adaptation into their own hands through DIY greening and nature-based solutions. Citizen-driven greening efforts can be more approachable, inclusive, and flexible than government-driven projects (Brink and Wamsler 2018; Brody et al. 2018). Urban residents benefit from the proximity to the local reality which leads to small-scale and tangible interventions, demonstrating the power of informal collective modes of action (Cloutier et al. 2018). While both CBA and DIY approaches can bring diverse stakeholders together in a way that is more collaborative than confrontational, they bring with them a risk of masking the lack of local institutional capacity and action towards urban adaptation (Anguelovski and Carmin 2011; Cloutier et al. 2018). Nonetheless, CBA and DIY initiatives that focus on climate vulnerability and bottom-up approaches can pioneer new practices, innovate, and act independently from government and have the potential to challenge the status quo and business as usual policies (Frantzeskaki et al. 2018). In doing so, they can be *a beacon of hope in this uncertain world* (Shammin et al. 2022).

An increase in co-production and participatory practices when planning for climate change adaptation is argued to increase community resilience (Mees et al. 2019; Robertson et al. 2021). A climate resilient community is one that can prepare for climate-related threats, reduce risk, recover from disaster, adapt to environmental changes, and preserve the health and culture of the community (Brody et al. 2018; Abrash Walton et al. 2021). Community resilience is often the result of bottom-up community initiatives and organisational networks and can come from formal or informal processes and actors (Robertson et al. 2021).

Community groups, non-profits, advocates, and neighbours are all an essential part of the climate adaptation process. From advocacy, to planning, to implementation, bringing in the people who are impacted the most is a key part of successful and equitable climate action. Furthermore, these groups and individuals working together towards just and sustainable futures are central to the debate around transformative adaptation. The section that follows summarises the growing conversation amongst scholars and practitioners around the role of participatory, collective, and community-driven adaptation in realising a future for cities that is inclusive and transformative.

2.4.4 The Transformative Potential of Participatory and Community-Driven Urban Climate Adaptation

There is a growing literature on what role community groups can play in promoting plans and projects that reflect the needs of the community and further equitable and just adaptation (Frantzeskaki et al. 2018). Community-based organisations and civil society are changing deep-rooted values and beliefs in cities with visions of more sustainable and equitable futures (Moore and Westley 2011; Frantzeskaki et al. 2018). In addition to CBA initiatives and social and environmental movements, public participation in adaptation planning can also lead to transformative action when (1) there is recognition of all actors, (2) meaningful engagement in all decision-making stages, and (3) full community decision-making power of those involved (Cattino and Reckien 2021). Community actions and movements, such as coalition building, are powerful in that they demonstrate how to build a political movement for just adaptation (Shi and Moser 2021). There is a need to look to both the collective and individual as key actors in intentional transformational adaptation. Organisations, networks, and coalitions can help to envision alternative futures, create and maintain momentum, and hold space for experimentation (Moore et al. 2014; Moore and Milkoreit 2020). It is this people-centric and participatory approach that can bring about systemic change through bottom-up mobilisation and alternative approaches (Revi et al. 2020) opening new avenues of transformative adaptation that fully embraces the principles of climate justice.

2.5 Conclusion

To plan for and act towards equitable and sustainable futures, decision makers, practitioners, and scholars must engage with alternative perspectives, knowledge, and cultures. This chapter strives to do just that; outline evolving debates around urban climate justice and transformative adaptation and explore their commonalities and contributions. To uncover and untangle the complex strategies and processes behind equitable climate adaptation and climate resilience in urban areas, this chapter examines, from a theoretical perspective, the role of civil society in climate planning and action, and more concretely community-driven climate adaptation, and discusses their transformative potential.

We have initiated the chapter with a review of climate governance and the importance of cities when preparing for climate change. Subsequently we have followed the origins and evolution of climate justice and the call for transformative and equitable climate adaptation in cities. A review of the literature on transformative adaptation has highlighted the importance of participatory and inclusive processes. Therefore, the chapter has looked at how civil society and community-based organisations engage in urban climate adaptation. We argue that communities need to be included in resilience and adaptation planning if a transformative change that embraces the principles of climate justice is to take place.

Acknowledgements This research has received funding from the Spanish Ministry of Science through the Plan Nacional RESCITIES (PGC2018-100996-AI00(MCIU/AEI/FEDER, UE)). K. Strange acknowledges grant funding from the Universitat Oberta de Catalunya. M. Satorras acknowledges the 'Juan de la Cierva – Formación' research fellowship (FJCI-2017-31723) funded by the Spanish Research Agency.

References

Abrash WA, Marr J, Cahillane MJ, Bush K (2021) Building Community Resilience to Disasters: A Review of Interventions to Improve and Measure Public Health Outcomes in the Northeastern United States. Sustainability 13(21):11699. https://doi.org/10.3390/su132111699

Agarwal A, Nurain S (1991) Global warming in an unequal world. Center for Science and the Environment, New Delhi

Ajibade I, Adams EA (2019) Planning principles and assessment of transformational adaptation: towards a refined ethical approach. Clim Dev 11(10):850–862. https://doi.org/10.1080/17565529.2019.1580557

Angelo H, Wachsmuth D (2015) Urbanizing urban political ecology: a critique of methodological cityism. Int J Urban Reg Res 39(1):16–27

Anguelovski I, Carmin JA (2011) Something borrowed, everything new: innovation and institutionalization in urban climate governance. Curr Opin Environ Sustain 3(3):169–175

Anguelovski I, Shi L, Chu E, Gallagher D, Goh K, Lamb Z, Reeve K, Teicher H (2016) Equity impacts of urban land use planning for climate adaptation: critical perspectives from the global north and south. J Plan Educ Res 36(3):333–348

Anguelovski I, Connolly JJT, Pearsall H, Shokry G, Checker M, Maantay J, Gould K, Lewis T, Maroko A, Roberts JT (2019) Why green "climate gentrification" threatens poor and vulnerable populations. PNAS 116(52):26139–26143. https://doi.org/10.1073/pnas.1920490117

Anguelovski I, Brand AL, Connolly JJT, Corbera E, Kotsila P, Steil J, Garcia-Lamarca M, Triguero-Mas M, Cole H, Baró F, Langemeyer J, Pérez del Pulgar C, Shokry G, Sekulova F, Argüelles L (2020) Expanding the boundaries of justice in urban greening scholarship: toward an emancipatory, antisubordination, intersectional, and relational approach. Ann Am Assoc Geogr 110(6):1743–1769. https://doi.org/10.1080/24694452.2020.1740579

Archer D, Almansi F, DiGregorio M, Roberts D, Sharma D, Syam D (2014) Moving towards inclusive urban adaptation: approaches to integrating community-based adaptation to climate change at city and national scale. Clim Dev 6(4):345–356. https://doi.org/10.1080/17565529.2014.918868

Arikan Y (2021) COP26 outcomes: multilevel action is the beacon of hope to keep the 1.5 degree goal alive. CityTalk, a blog by ICLEI. https://talkofthecities.iclei.org/cop26-outcomes-multilevel-action-is-the-beacon-of-hope-to-keep-the-1-5-degree-goal-alive/

Arnstein SR (1969) A ladder of citizen participation. J Am Plan Assoc 35(4):216–224

Aylett A (2015) Institutionalizing the urban governance of climate change adaptation: Results of an international survey. Urban Climate 144–16. S2212095515300031 https://doi.org/10.1016/j.uclim.2015.06.005

Bassett TJ, Fogelman C (2013) Déjà vu or something new? The adaptation concept in the climate change literature. Geoforum 4842–53. S0016718513000821 https://doi.org/10.1016/j.geoforum.2013.04.010

Bazaz A, Bertoldi P, Buckeridge M et al (2018) Summary for urban policymakers: what the IPCC Special Report on global warming of 1.5°C means for cities. IHHS Indian Institute for Human Settlements, Bengaluru
Blythe J, Silver J, Louisa AD, Bennett N, Moore M, Morrison TH, Brown K (2018) The dark side of transformation: latent risks in contemporary sustainability discourse. Antipode 50(5):1206–1223
Boyd E, Folke C (2011) Adapting institutions: governance, complexity and social–ecological resilience. Cambridge University Press, Cambridge
Boyd E, Juhola S (2015) Adaptive climate change governance for urban resilience. Urban Stud 52(7):1234–1264
Brink E, Wamsler C (2018) Collaborative governance for climate change adaptation: mapping citizen–municipality interactions. Environ Policy Gov 28(2):82–97
Brody L, Chelleri L, Baró F, Ruiz-Mallen I (2018) Enhancing community resilience in Barcelona: addressing climate change and social justice through spaces of comanagement. In: Galderisis A, Colucci A (eds) Smart, resilient and transition cities: emerging approaches and tools for a climate-sensitive urban development. Elsevier, Amsterdam
Bryant B, Mohai P (1992) Race and the incidence of environmental hazards: a time for discourse. Westview Press, Boulder
Bulkeley H (2021) Climate changed urban futures: environmental politics in the anthropocene city. Environ Polit 30(1–2):266–284. https://doi.org/10.1080/09644016.2021.1880713
Bulkeley H, Castán Broto V (2013) Government by experiment? Global cities and the governing of climate change. Trans Inst Br Geogr 38(3) 361–375. https://doi.org/10.1111/j.1475-5661.2012.00535.x
Bulkeley H, Carmin JA, Castán Broto V, Edwards GAS, Fuller S (2013) Climate justice and global cities: mapping the emerging discourses. Glob Environ Chang 23(5):914–925
Busch H, Bendlin L, Fenton P (2018) Shaping local response – the influence of transnational municipal climate networks on urban climate governance. Urban Clim 24:221–230. https://doi.org/10.1016/j.uclim.2018.03.004
Caniglia BS, Brulle RJ, Szasz A (2015) Civil Society, Social Movements, and Climate Change. In Climate Change and Society: Sociological Perspectives (eds. Dunlap RE and Brulle RJ). Oxford Scholarship Online
C40 (2020) About. https://www.c40.org/about. Accessed 26 Mar 2020
Castán Broto V (2017) Urban governance and the politics of climate change. World Dev 93:1–15. https://doi.org/10.1016/j.worlddev.2016.12.03
Castán Broto V, Bulkeley H (2013) A survey of urban climate change experiments in 100 cities. Glob Environ Chang 23(1):92–102
Castán Broto V, Robin E (2021) Climate urbanism as critical urban theory. Urban Geogr 42(6):715–720. https://doi.org/10.1080/02723638.2020.1850617
Castán Broto V, Westman LK (2020) Ten years after Copenhagen: reimagining climate change governance in urban areas. WIREs Clim Change 11(4):e643. https://doi.org/10.1002/wcc.643
Cattino M, Reckien D (2021) Does public participation lead to more ambitious and transformative local climate change planning? Curr Opin Environ Sustain 52:100–110. https://doi.org/10.1016/j.cosust.2021.03.013
Chu E, Anguelovski I, Roberts D (2017) Climate adaptation as strategic urbanism: assessing opportunities and uncertainties for equity and inclusive development in cities. Cities 60:378–387
Chu E, Brown A, Michael K, Du J, Lwasa S, Mahendra A (2019) Unlocking the potential for transformative climate adaptation in cities. Background paper prepared for the Global Commission on Adaptation, Washington, DC/Rotterdam
Cloutier G, Papin M, Bizier C (2018) Do-it-yourself (DIY) adaptation: civic initiatives as drivers to address climate change at the urban scale. Cities 74:284–291
Coggins S, Berrang-Ford L, Hyams K et al (2021) Empirical assessment of equity and justice in climate adaptation literature: a systematic map. Environ Res Lett 16:073003

Cook IR, Swyngedouw E (2012) Cities, social cohesion and the environment: towards a future research agenda. Urban Stud 49(9):1959–1979. https://doi.org/10.1177/0042098012444887
COP26 (2021) Delivering the Glasgow Climate Pact. https://ukcop26.org/
Dawson A (2017) Extreme Cities The peril and promise of urban life in the age of climate change. Verso, London and New York
de Moor J, De Vydt M, Uba K, Wahlström M (2021) New kids on the block: taking stock of the recent cycle of climate activism. Soc Mov Stud 20(5):619–625. https://doi.org/10.1080/14742837.2020.1836617
der Heijden J (2019) Studying urban climate governance: where to begin, what to look for, and how to make a meaningful contribution to scholarship and practice. Earth Syst Gov 1:100005. https://doi.org/10.1016/j.esg.2019.100005
Deubelli TM, Mechler R (2021) Perspectives on transformational change in climate risk management and adaptation. Environ Res Lett 16:053002. https://doi.org/10.1088/1748-9326/abd42d
Dodman D, Mitlin D (2013) Challenges for community-based adaptation: discovering the potential for transformation. J Int Dev 25(5):640–659
Elmqvist T, Andersson E, Frantzeskaki N et al (2019) Sustainability and resilience for transformation in the urban century. Nat Sustain 2(4):267–273
Eriksen SH, Nightingale AJ, Eakin H (2015) Reframing adaptation: the political nature of climate change adaptation. Glob Environ Chang 35:523–533
Fainstein SS (2018) Resilience and justice: planning for New York City. Urban Geogr 39(8):1268–1275. https://doi.org/10.1080/02723638.2018.1448571
Fedele G, Donatti CI, Harvey CA, Hannah L, Hole DG (2019) Transformative adaptation to climate change for sustainable social-ecological systems. Environ Sci & Policy 101116–125 S1462901119305337 https://doi.org/10.1016/j.envsci.2019.07.001
Few R, Brown K, Tompkins EL (2007) Public participation and climate change adaptation: avoiding the illusion of inclusion. Clim Pol 7(1):46–59
Fiack D, Cumberbatch J, Sutherland M, Zerphey N (2021) Sustainable adaptation: social equity and local climate adaptation planning in U.S. cities. Cities 115:103235
Fisher D (2019) The broader importance of #FridaysForFuture. Nat Clim Chang 9:428–433
Forsyth T (2013) Community-based adaptation: a review of past and future challenges. WIREs Clim Change 4:439–446
Frantzeskaki N, Dumitru A, Wittmayer J, Avelino F, Moore M-L (2018) To transform cities, support civil society. Urban Planet:281–302. https://doi.org/10.1017/9781316647554.016
Fuhr H, Hickmann T, Kern K (2018) The role of cities in multi-level climate governance: local climate policies and the 1.5°C target. Curr Opin Environ Sustain 30:1–6. https://doi.org/10.1016/j.cosust.2017.10.006
Global Covenant of Mayors (2020) About. https://www.globalcovenantofmayors.org/about/
Gordon DJ (2018) Global urban climate governance in three and a half parts: experimentation, coordination, integration (and contestation). WIREs Clim Change 9(6):1–15. https://doi.org/10.1002/wcc.546
Haasnoot M, Kwakkel JH, Walker WE, ter Maat J (2013) Dynamic adaptive policy pathways: a method for crafting robust decisions for a deeply uncertain world. Glob Environ Chang 23(2):485–498
Hamel P, Hamann M, Kuiper JJ, Andersson E, Arkema KK, Silver JM, Daily GC, Guerry AD (2021) Blending ecosystem service and resilience perspectives in planning of natural infrastructure: lessons from the San Francisco Bay Area. Front Environ Sci 9:1–13. https://doi.org/10.3389/fenvs.2021.601136
Hoffman JS, Shandas V, Pendleton N (2020) The effects of historical housing policies on resident exposure to intra-urban heat: a study of 108 US urban areas. Climate 8(1):1–15. https://doi.org/10.3390/cli8010012
Holland B (2017) Procedural justice in local climate adaptation: political capabilities and transformational change. Environ Pol 26(3):391–412 2 https://doi.org/10.1080/09644016.2017.1287625

Hölscher K, Frantzeskaki N, McPhearson T, Loorbach D (2019a) Tales of transforming cities: transformative climate governance capacities in New York City, U.S. and Rotterdam, Netherlands. J Environ Manage 231:843–857
Hölscher K, Frantzeskaki N, McPhearson T, Loorbach D (2019b) Capacities for urban transformations governance and the case of New York City. Cities 94:186–199
Hughes S (2020) Principles, drivers, and policy tools for just climate change adaptation in legacy cities. Environ Sci Policy 111:35–41. https://doi.org/10.1016/j.envsci.2020.05.007
Hughes S, Hoffmann M (2020) Just urban transitions: toward a research agenda. WIREs Clim Change 11(3):1–11. https://doi.org/10.1002/wcc.640
Hurlimann AC, Moosavi S, Browne GR (2021) Climate change transformation: a definition and typology to guide decision making in urban environments. Sustain Cities Soc 70:102890. https://doi.org/10.1016/j.scs.2021.102890
IPCC (2012) Managing the risks of extreme events and disasters to advance climate change adaptation. A special report of Working Groups I and II of the Intergovernmental Panel on Climate Change. Cambridge University Press, Cambridge/New York
IPCC (2014) Urban areas. In: Field LLW, Barros VR, Dokken DJ, Mach KJ, Mastrandrea MD, Bilir TE, Chatterjee M, Ebi KL, Estrada YO, Genova RC, Girma B, Kissel ES, Levy AN, MacCracken S, Mastrandrea PR (eds) Contribution of Working Group II to the fifth assessment report of the Intergovernmental Panel on Climate Change. Cambridge University Press, Cambridge
IPCC (2021) Summary for policymakers. In: Climate change 2021: the physical science basis. Contribution of Working Group I to the sixth assessment report of the intergovernmental panel on climate change. Cambridge University Press, Cambridge
Iwaniec DM, Cook EM, Barbosa O, Grimm NB (2019) The framing of urban sustainability transformations. Sustainability 11(3):1–10. https://doi.org/10.3390/su11030573
Jamieson D (2001) Climate change and global environmental justice. In: Miller CA, Edwards PN (eds) Changing the atmosphere: expert knowledge and environmental governance. MIT Press, Cambridge, pp 287–308
Jamison A (2010) Climate change knowledge and social movement theory. WIREs Clim Change 1(6):811–823 https://doi.org/10.1002/wcc.88
Kaika M (2017) 'Don't call me resilient again!': the New Urban Agenda as immunology … or … what happens when communities refuse to be vaccinated with 'smart cities' and indicators. Environ Urban 29(1):89–102
Kates RW, Wilbanks TJ (2012) Transformational adaptation when incremental adaptations to climate change are insufficient. Proc Natl Acad Sci 109(19):7156–7161 https://doi.org/10.1073/pnas.1115521109
Kauffman N, Hill K (2021) Climate change, adaptation planning and institutional integration: a literature review and framework. Sustainability 13(19):10708. https://doi.org/10.3390/su131910708
Klein J, Araos M, Karimo A, Heikkinen M, Ylä-Anttila T, Juhola S (2018) The role of the private sector and citizens in urban climate change adaptation: evidence from a global assessment of large cities. Glob Environ Chang 53:127–136. https://doi.org/10.1016/j.gloenvcha.2018.09.012
Kuhl L, Rahman MF, McCraine S, Krause D, Hossain MF, Bahadur AV, Huq S (2021) Transformational Adaptation in the Context of Coastal Cities. Annu Rev Environ Resour 46(1):449–479 https://doi.org/10.1146/annurev-environ-012420-045211
LGMA (2021) LGMA Glasgow #TIME4MULTILEVELACTION Roadmap. Available at: https://www.cities-and-regions.org/wp-content/uploads/lgma_glasgow_time4multilevelactionroadmap_final.pdf
Long J, Rice JL (2019) From sustainable urbanism to climate urbanism. Urban Stud 56(5):992–1008. https://doi.org/10.1177/0042098018770846
Malloy JT, Ashcraft CM (2020) A framework for implementing socially just climate adaptation. Clim Chang 160:1–14. https://doi.org/10.1007/s10584-020-02705-6

Manuel-Navarrete D, Pelling M (2015) Subjectivity and the politics of transformation in response to development and environmental change. Glob Environ Chang 35:558–569
Maru YT, Stafford Smith M (2014) Reframing adaptation pathways. Glob Environ Chang 28:322–324
Matin N, Forrester J, Ensor J (2018) What is equitable resilience? World Dev 109:197–205
McPhearson T (2020) Transforming cities and science for climate change resilience in the Anthropocene. In: Hölscher K, Frantzeskaki N (eds) Transformative climate governance. Palgrave studies in environmental transformation, transition and accountability. Palgrave Macmillan, Cham, pp 99–111
Meerow S, Woodruff S (2020) Seven principles of strong climate change planning. J Am Plan Assoc 86(1):39–46. https://doi.org/10.1080/01944363.2019.1652108
Meerow S, Newell JP, Stults M (2016) Defining urban resilience: a review. Landsc Urban Plan 147:38–49
Meerow S, Pajouhesh P, Miller TR (2019) Social equity in urban resilience planning. Local Environ 24(9):793–808. https://doi.org/10.1080/13549839.2019.1645103
Mees HLP, Uittenbroek CJ, Hegger DLT, Driessen PPJ (2019) From citizen participation to government participation: an exploration of the roles of local governments in community initiatives for climate change adaptation in the Netherlands. Environ Policy Gov 29(3):198–208
Mikulewicz M (2019) Thwarting adaptation's potential? A critique of resilience and climate-resilient development. Geoforum 104:267–282
Mix TL (2011) Rally the People: Building Local-Environmental Justice Grassroots Coalitions and Enhancing Social Capital*. Sociol Inq 81(2):174–194. https://doi.org/10.1111/j.1475-682X.2011.00367.x
Mohtat N, Khirfan L (2021) The climate justice pillars vis-a-vis urban form adaptation to climate change: a review. Urban Clim 39:100951. https://doi.org/10.1016/j.uclim.2021.100951
Moore M-L, Milkoreit M (2020) Imagination and transformations to sustainable and just futures. Elem Sci Anth 8:1. https://doi.org/10.1525/elementa.2020.081
Moore M-L and Westley FR (2011) 'Public Sector Policy and Strategies for Facilitating Social Innovation', Horizons, 11. Online: http://www.policyresearch.gc.ca/page.asp?pagenm=pub_index. Accessed: 17 November 2011.
Moore M-L, Tjornbo O, Enfors E, Knapp C, Hodbod J, Baggio JA, Norström A, Olsson P, Biggs D (2014) Studying the complexity of change: toward an analytical framework for understanding deliberate social-ecological transformations. Ecol Soc 19(4):54 https://doi.org/10.5751/ES-06966-190454
Moser SC, Pike C (2015) Community engagement on adaptation: meeting a growing capacity need. Urban Clim 14:111–115
Newell P, Srivastava S, Naess LO, Contreras GT, Price R (2021) Toward transformative climate justice: an emerging research agenda. WIREs Clim Change 12(6). https://doi.org/10.1002/wcc.733
Nightingale AJ (2017) Power and politics in climate change adaptation efforts: Struggles over authority and recognition in the context of political instability. Geoforum 8411–20 S001671851730129X https://doi.org/10.1016/j.geoforum.2017.05.011
Nordgren J, Stults M, Meerow S (2016) Supporting local climate change adaptation: where we are and where we need to go. Environ Sci Policy 66:344–352
Pelling M (2011) Adaptation to climate change: from resilience to transformation. Routledge, New York
Pelling M, O'Brien K, Matyas D (2015) Adaptation and transformation. Clim Chang 133(1):113–127 https://doi.org/10.1007/s10584-014-1303-0
Pellow DN (2001). Environmental Justice and the Political Process: Movements, Corporations, and the State. Sociol Q 42(1):47–67. http://www.jstor.org/stable/4120925
Reckien D, Creutzig F, Fernandez B, Lwasa S, Tovar-Restrepo M, Mcevoy D, Satterthwaite D (2017) Climate change, equity and the sustainable development goals: an urban perspective. Environ Urban 29(1):159–182 https://doi.org/10.1177/0956247816677778
Resilient Cities (2020) Our impact. https://www.100resilientcities.org/our-impact/

Revi A, Satterthwaite D, Aragón-Durand F et al (2014) Towards transformative adaptation in cities: the IPCC's Fifth Assessment. Environ Urban 26(1):11–28. https://doi.org/10.1177/0956247814523539
Revi A, Anguelovski I, Filho WL, Olazabal M, Chu E, Cooper JT, Garschagen M, Nelson DR (2020) Transformative adaptation in cities. One Earth 3(4):384–387. https://doi.org/10.1016/j.oneear.2020.10.002
Robertson T, Docherty P, Millar F et al (2021) Theory and practice of building community resilience to extreme events. Int J Disaster Risk Reduct 59:102253. https://doi.org/10.1016/j.ijdrr.2021.102253
Romero-Lankao P, Bulkeley H, Pelling M, Burch S, Gordon DJ, Gupta J, Johnson C, Kurian P, Lecavalier E, Simon D, Tozer L, Ziervogel G, Munshi D (2018) Urban transformative potential in a changing climate. Nat Clim Chang 8:754–756
Rosenzweig C, Solecki W, Romero-Lankao P, Mehrotra S, Dhakal S, Bowman T, Ali Ibrahim S (2015) ARC3.2 summary for city leaders. Urban Climate Change Research Network. Columbia University, New York
Ruiz-Mallén I (2020) Co-production and resilient cities to climate change. In: Nared J, Bole D (eds) The participatory research and planning in practice. The urban book series. Springer, Cham
Ribot J (2014) Cause and response: vulnerability and climate in the Anthropocene. The Journal of Peasant Studies 41(5):667–705. https://doi.org/10.1080/03066150.2014.894911
Sarzynski A (2015) Public participation, civic capacity, and climate change adaptation in cities. Urban Clim 14:52–67
Satorras M (2021) From commissioned reports to co-production processes: evolution of urban climate policy-making through the case of Barcelona. In: Leal Filho W, Luetz JM, Ayal D (eds) Handbook of climate change management: research, leadership, transformation. Springer, Cham, pp 4379–4401. https://doi.org/10.1007/978-3-030-57281-5_42
Satorras M, Ruiz-Mallén I, Monterde A, March H (2020) Co-production of urban climate planning: insights from the Barcelona climate plan. Cities 106:102887. https://doi.org/10.1016/j.cities.2020.102887
Schinko T (2020) Overcoming political climate-change apathy in the era of #FridaysForFuture. One Earth 2(1):20–23. https://doi.org/10.1016/j.oneear.2019.12.012
Schipper L (2006) Conceptual History of Adaptation in the UNFCCC Process. Review of European Community and International Environmental Law 15(1):82–92 https://doi.org/10.1111/j.1467-9388.2006.00501.x
Schlosberg D (2012) Climate justice and capabilities: a framework for adaptation policy. Ethics Int Aff 26(4):445–461
Schlosberg D (2013) Theorising environmental justice: the expanding sphere of a discourse. Env Polit 22(1):37–55
Schlosberg D, Collins LB, Niemeyer S (2017) Adaptation policy and community discourse: risk vulnerability and just transformation. Environ Pol 26(3):413–437 https://doi.org/10.1080/09644016.2017.1287628
Sengers F, Turnheim B, Berkhout F (2021) Beyond experiments: embedding outcomes in climate governance. Environ Plan C: Politics Space 39(6):1148–1171. https://doi.org/10.1177/2399654420953861
Shammin MR, Enamul Haque AK, Faisal IM (2022) A framework for climate resilient community-based adaptation. In: Enamul Haque A, Mukhopadhyay P, Nepal M, Shammin MR (eds) Climate change and community resilience. Springer, Singapore. https://doi.org/10.1007/978-981-16-0680-9_2
Shi L, Moser S (2021) Transformative climate adaptation in the United States: trends and prospects. Science 372(6549):1085–1086. https://doi.org/10.1126/science.abc8054
Shue H (1993) Subsistence emissions and luxury emissions. Law Policy 15(1):39–59
Singer P (2004) One world: the ethics of globalization. Yale University Press, New Haven
Sultana F (2021) Political ecology II: conjunctures, crises, and critical publics. Prog Hum Geogr 45(6):1721–1730. https://doi.org/10.1177/03091325211028665
Svarstad H, Benjaminsen TA (2020) Reading radical environmental justice through a political ecology lens. Geoforum 108:1–11

Swyngedouw E, Heynen NC (2003) Urban political ecology, justice and the politics of scale. Antipode 35(5):898–918
UN-Habitat (2011) Cities and climate change: global report on human settlements 2011. Earthscan, London
UNISDR (2015) Sendai framework for disaster risk reduction 2015–2030. United Nations
United Nations (UN) (2019) World urbanization prospects: the 2018 revision. Department of Economic and Social Affairs, Population Division. United Nations, New York
United Nations General Assembly (2015) Draft outcome document of the United Nations summit for the adoption of the post 2015 development agenda. Sixty-ninth session A/69/L.85, 12 August
Wamsler C (2017) Stakeholder involvement in strategic adaptation planning: transdisciplinarity and co-production at stake? Environ Sci Policy 75:148–157
Wardekker A (2021) Contrasting the framing of urban climate resilience. Sustain Cities Soc 75:103258. https://doi.org/10.1016/j.scs.2021.103258
Whitehead M (2013) Neoliberal urban environmentalism and the adaptive city: towards a critical urban theory and climate change. Urban Stud 50(7):1348–1367
Wilson RS, Herziger A, Hamilton M, Brooks JS (2020) From incremental to transformative adaptation in individual responses to climate-exacerbated hazards. Nat Clim Chang 10(3):200–208. https://doi.org/10.1038/s41558-020-0691-6
Wise RM, Fazey I, Stafford Smith M, Park SE, Eakin HC, Archer Van Garderen ERM, Campbell B (2014) Reconceptualising adaptation to climate change as part of pathways of change and response. Glob Environ Chang 28:325–336
Wolfram M, Frantzeskaki N, Maschmeyer S (2016) Cities, systems and sustainability: status and perspectives of research on urban transformations. Curr Opin Environ Sustain 22:18–25
Woodruff S C, Meerow S, Stults M, Wilkins C (2018) Adaptation to resilience planning: alternative pathways to prepare for climate change. J Plan Educ Res. Article 0739456X1880105. https://doi.org/10.1177/0739456X18801057
World Climate Research Programme (2019) Global research and action agenda on cities and climate change science. WCRP Report No. 3/2019
Woroniecki S, Krüger R, Rau AL et al (2019) The framing of power in climate change adaptation research. WIREs Clim Change 10(6):1–21. https://doi.org/10.1002/wcc.617
Ziervogel G, Pelling M, Cartwright A, Chu E, Deshpande T, Harris L, Hyams K, Kaunda J, Klaus B, Michael K, Pasquini L, Pharoah R, Rodina L, Scott D, Zweig P (2017) Inserting rights and justice into urban resilience: a focus on everyday risk. Environ Urban 29(1):123–138
Zografos C, Robbins P (2020) Green sacrifice zones, or why a green new deal cannot ignore the cost shifts of just transitions. One Earth 3(5):543–546. https://doi.org/10.1016/j.oneear.2020.10.012
Zografos C, Klause KA, Connolly JJT, Anguelovski I (2020) The everyday politics of urban transformational adaptation: struggles for authority and the Barcelona superblock project. Cities 99:102613

Part II
Uneven Implications of Top-Down Resilience

Chapter 3
Urban Resilience in Perspective: Tracing the Origins and Evolution of Urban Green Spaces in Barcelona

Jordi Honey-Rosés

Abstract The transition to the resilient city would benefit from an appreciation of our urban history and the challenges confronted by city builders of the past. This historical lens may provide contemporary planners with greater perspective, context and humility when envisioning the transformative changes needed to address our climate emergency. A historical perspective is particularly useful when examining parks and urban green spaces, which are key features of the resilient city because they provide essential ecosystem services related to biodiversity, water flows, air pollution, temperature control, health and recreation. This chapter traces the historical planning decisions that led to the configuration of the current green network in Barcelona, Spain. I propose a historical typology of urban green space that illustrates the origins and causes of its deficient network. Contemporary planners in Barcelona have now set ambitious goals to green the city with new street designs and urban greening projects. However these greening efforts have been criticised for their unintended consequences, as they bring in more visitors and may accelerate gentrification. Looking ahead, the pandemic may re-calibrate our expectations about urban green space, especially with regard to how much, how close and how crowded they should be.

Keywords Barcelona · Green space · Resilience planning · Urban history

3.1 Introduction

The literature on urban resilience is packed with urgency. No drama is spared when calling for action to address our environmental crisis (Rockström et al. 2009). Local governments have declared a climate emergency (Gills and Morgan 2020), and this

J. Honey-Rosés (✉)
Institut de Ciència i Tecnologia Ambientals, Universitat Autònoma de Barcelona, Cerdanyola del Vallès, Spain
e-mail: jordi.honey@uab.cat

I. Ruiz-Mallén et al. (eds.), *Urban Resilience to the Climate Emergency*, The Urban Book Series, https://doi.org/10.1007/978-3-031-07301-4_3

emergency requires unprecedented transformations in our energy system, land use, mobility networks, and governance structures (Satterthwaite and Dodman 2013; Satorras et al. 2020). We have entered a new geological era, the *Anthropocence*, characterised by unprecedented planetary change driven by human alterations to our Earth's support systems (Lewis and Maslin 2015). To sustain urban life as we know it, cities must undergo a profound re-organisation and transformation, and failure to make these changes may bring human society to irreversible collapse (Servigne and Stevens 2020).

The COVID-19 pandemic of 2020–2021 has forced us envision radical new global futures and stretched our imagination. We have gained a collective appreciation of exponential growth, our individual and social vulnerability and global interdependency. Now more than ever, the notion of global collapse seems plausible. For decades scholars have been discussing the limits to economic growth and the risk of ecological collapse (Rees and Wackernagel 1996; Kallis 2011), although it is unclear if this moment will produce significant shifts in the dominant production and consumption systems that have brought us here. Clearly, the state of our planet merits re-orientation. It would be irresponsible to ignore the extreme environmental pressures that we currently confront.

Yet, as we discuss urban resilience and work to transition to more climate resilient cities, we would benefit from looking back at our urban history to appreciate the transformations led by generations that came before us (Grau-Satorras et al. 2016; Gorostiza et al. 2021). Urban planners from the late nineteenth century also confronted an unprecedented public health crisis, swift technological change, social upheaval and even revolutionary violence. Contextualizing our current challenges may help clarify that our generation is certainly not the first to call for ambitious urban transformations. In many instances, past transformations were in fact substantially more radical that what progressive voices are calling for today. Compare for instance, the demand for new bicycle lanes with the construction of an underground sanitary infrastructure. Furthermore, city planners in the late nineteenth century operated under the pressure of social unrest or revolutionary violence that questioned fundamental governance structures and political systems. In comparison, our political systems are highly stable.

If we look back into our urban history, what can resilience planners learn from past transformations? How might a greater appreciation of urban history inform current discussions on urban climate resilience? And to what extent are current urban transitions constrained by the historic evolution of the city, especially with regard to land use and green space planning? This chapter explores these questions by tracing the evolution of green space planning in Barcelona, Spain. I argue that a historical lens may provide contemporary planners and scholars with greater perspective, context and humility when envisioning the transformative changes required to achieve the resilient city. A historical analysis may also explain how a city has inherited its current configuration and morphology. Appreciation of the origins of these constraints may offer clues about how to create a more resilient green space network.

Environmental historians already contribute to climate change research (Carey and Garone 2014) and in particular, research on climate adaptations (Wise et al. 2014; Adamson et al. 2018) in which scholars are uncovering past examples of how communities have responded to stressors, developed coping mechanisms, and climate related adaptations (Grau-Satorras et al. 2016). For instance, a review of sanitary infrastructure in Zurich illustrates that city managers are continuously confronted with challenges that the physical infrastructure was not designed for, thereby forcing innovation and adaptions for successful operation in novel and unexpected conditions (Neumann et al. 2015). The long-term view helps contextualise the challenges confronted by planners and city managers today.

My historic review focuses on urban green spaces, which are key features of the climate resilient city because they provide essential ecosystem services related to biodiversity, water flows, air pollution, temperature control, health and recreation. The spatial distribution of these green spaces will impact community vulnerabilities by neighbourhood. I aim to make my argument applicable to cities generally, although the lessons are derived from the history of green space planning in Barcelona. By reviewing secondary sources and planning documents, I first present an account of the origins of Barcelona's greenspace network. Then, I apply this historical lens to develop a typology of urban green space for the city that illustrates the origins (and causes) of its deficient green space network. A historic view allows us to see that the current inequities in access to green spaces are the result of past planning decisions. Barcelona is a special case for urban planners because its orthogonal city plan was designed by the first urban planner in the modern sense, Ildefons Cerdà, who prioritised public health objectives and the creation of open space in his innovative city extension plan. In practice, however, the plan was not implemented as envisioned, creating a dense city with limited green space. After years of neglect, the city planners in the early democratic period (1978–1982) purchased private lands that were converted to green spaces in lower income neighbourhoods. While the parks created during this period met social justice objectives, more recent city greening initiatives have been framed as part of Barcelona City Council's strategy to become a more competitive and global city. Finally, I end with reflections on how the pandemic may re-calibrate our expectations about urban green spaces.

3.2 Urban Resilience in Perspective

3.2.1 The Novelty of the Nineteenth Century Urban Challenges

It is worth considering what it must have been like to live in an industrializing city such as London, Manchester, Paris, or New York in 1850. Migrants travelled to the city to find overcrowded housing, strangers in the streets, air filled with soot, and sewage, filth and grime on every surface of the built environment. Workers suffered from chronic malnourishment, infant mortality was high, and the water was polluted. Standing from our place in history today, it is easy to forget that the experience of an industrial worker during this period was entirely new, unseen before, and the product of the novel process of urbanisation itself.

Disease and plague ravished the cities. The cholera outbreaks of 1832 and 1849 in London killed 14,137 people (Johnson 2006). Furthermore, the best medical scientists were convinced that cholera was transmitted through the stench in the air, known as miasma theory, leading to misguided protection efforts. It was not until John Snow identified the origins of cholera as a water-borne disease in 1854, that the miasma theory began to be put to rest. And even after having understood the leading scientific causes of waterborne diseases, it still took sanitarians decades to re-direct the flows of waste and water to protect public health.

Consider also, what it must have felt like to be responsible for managing the city during this period. Rights of way on streets remained undefined, ambiguous and leading to constant conflict and injury. The streets were incapable of absorbing the sewage – the novel product produced by mixing water with household waste. The filth on the street was unprecedented. New and untested technologies and infrastructure were needed to reform the city. Consider the scale of the proposal of creating a new sewerage system (Neumann et al. 2015). Urban planners of the nineteenth century confronted problems that were entirely new, unseen before, and required radical reconfiguration of urban space.

The dramatic language that circulates today to describe our current environmental crisis make it sound like previous generations have had it easy. Let us not forget the magnitude of the city building challenges from the nineteenth century. We are not the first to advocate for transformational changes to our city's infrastructure or urban metabolism. We should find at least some comfort knowing that previous generations succeeded in engineering healthier cities, and this feat was a major accomplishment of the nineteenth century city builders. Having briefly paused to consider the radical urban transformations led by generations before us, I now turn to the history of green spaces planning in Barcelona.

3.3 Tracing the Origins of Urban Green Spaces in Barcelona

The absence of open space and parks has been a recurring theme in Barcelona's urban history. For centuries, Barcelona was compressed behind medieval walls that squeezed out any opportunity for open space. In the mid 1800s Barcelona occupied the same urban footprint as it did in the 1300s, enclosed behind walls built in the fourteenth century. For 500 years, the population grew within the same land base, creating crowded neighbourhoods and compressed living conditions. Barcelona began to industrialise behind the medieval walls. Petitions to grow the city were repeatedly rejected by the ruling powers in Madrid. The lands surrounding the city were off limits to development under special military orders. Historic towns outside the city began to grow: *Sants, Sarrià, Sant Martí, Horta* and by the eighteenth century, *Gràcia*. But Barcelona remained enclosed. The walls became an odious symbol of the constraints and limitations imposed by distant rulers in Madrid (Hughes 1992). By the late 1850s Barcelona was one of the most densely populated cities in the world, with more than twice the urban density of Paris, and ten times the density of Victorian London[1] (Busquets 2005).

When the governing regime in Madrid finally gave Barcelona permission to grow outside its walls, city planners, architects and residents yearned to build a new and modern city, with wide streets, direct sunlight and ventilation (Solà-Morales 2008). They envisioned an extension plan that would connect the old city with the surrounding towns to make one urban continuum. They envisioned a cleaner, brighter and healthier city. The second half of the nineteenth century provided Barcelona with the chance to create a new city and transform it in a way that is hardly imaginable today.

In 1859, Barcelona city officials called for a design competition that would gather proposals for the new city extension (*Eixample*). The winning plan was submitted by the municipal architect, Antoni Rovira-i-Trias (1816–1886), and his proposal called for several large open spaces, parks, forests or gardens within the city fabric. Yet when the drawings of the Rovira Plan were sent to Madrid for final approval, the Spanish government overruled the decision, and informed city officials that the extension plan for Barcelona would be based on another submission, the one developed by Ildefons Cerdà (Busquets 2005). Why the central government overruled city officials remains a mystery, but this decision has had profound consequences for Barcelona, and for the field of urban planning.

Ildefons Cerdà (1815–1876) was the first urban planner in the modern sense. He is credited with having conceptualised the term *urbanisme* and thereby gave birth to city planning as a discipline (Magrinyà and Marzá 2017). If cities have existed for thousands of years, why was it that Ildefons Cerdà first brought the words urbanism and urbanist to print in 1859. Why then? In retrospect, we can see that the conditions of modern life demanded a distinctive understanding of the city (Sennett 2018). Urbanism weaves together architecture, design and engineering and Cerdà had the vision to anticipate this new conceptualization of the city.

Cerdà was a systems thinker, with a keen eye for the physical and social dimensions of cities. A civil engineer by training, he was the only participant in the design competition for Barcelona's extension plan without a degree in architecture. To prepare for his extension proposal, Cerdà went to extraordinary lengths to learn about the residents of the old city: the laborers, artisans and urban poor. He wanted to understand whom he was planning for, and aspired to develop a city that would improve their living conditions. He tabulated statistics on life expectancy, health, family size, and produced countless tables with descriptive statistics of Barcelona's residents. Cerdà also reached out to speak with residents, interviewing them in their

[1] Urban inhabitants per hectare London 86 p/ha (1858), Paris 384 p/ha (1859), Barcelona 859 p/ha (1859) (Busquets 2005).

homes, and aiming to gain a qualitative understanding of their hardship. It is likely that these conversations made a lasting impression on the young engineer, leading him to create a visionary plan, driven by egalitarian principles, that broke from the hierarchical tradition of the day (Hughes 1992).

Principles of equity permeated Cerdà's plan. He proposed that Barcelona should grow beyond the city walls in a grid form, with uniform blocks and chamfered corners, but development would only be built on two sides of each block, thereby creating large open spaces inside each grid cell. He also envisioned eight public parks, for a total of 82 ha and a large forest reserve near the *Besòs* river (Martí and Moreno 1974). His design was motivated by a desire to increase the flow of light, air and vehicles through city streets, all within a framework that would meet radical social objectives. Cerdà saw the grid morphology as a planners' best tool to create a more equitable society and ensure equal access to urban infrastructure, amenities and open space. The grid was simultaneously rational and a great equalizer, preventing neighbourhood distinctions by class. Some called his plan utopian (Hughes 1992). In this respect, Barcelona became one of the first large cities to translate utopian ideals into a city blueprint and built form. Cerdà's vision for an equitable distribution of green spaces ran contrary to the mainstream approach of his time, including the rival Plan Rovira which had proposed concentrating large open spaces in particular sections of the city hierarchy. Such a plan, had it been completed, would have produced inequitable greening by design, creating more desirable neighbourhoods and reinforcing class divisions. Cerdà aimed for the opposite and planned equity into the city, with open spaces for everyone.

Yet Cerdà's plan failed to live up to his grand vision for equality and open spaces for all. Development pressure led the city to approve development on all four sides of his planned blocks, thereby enclosing the grid and annulling his vision for an open city. Nor did the *Eixample* become the social equalizer by class or income. Those with the financial means to leave the cramped quarters of the old city were the first to go, leaving the urban poor behind. Even to this day, the old city is home to lower income households.

Today Cerdà's grid is the dominant urban feature in Barcelona, and this compact and dense urban form has produced the near opposite of what Cerdà envisioned, with most residents being of middle or high-income and having limited access to green space and parks. Several lower income neighbourhoods beyond the *Eixample* have greater access to green spaces and parks because of their location on the periphery. Therefore any discussion about green space planning, access and equity in Barcelona must be contextualised in the city's unique pattern of urban development: forced densification; the new grid as a utopian ideal; and the perversion of Cerdà's plan.

While it may be tempting to blame the Franco dictatorship (1939–1975) for the absence of green space in Barcelona, a careful look reveals that its origins go back further. Still, the fascist period was not kind to Barcelona, as the city fell into decay and became hostage to real-estate interests (Ynfante 1974). By the time Barcelona recovered its democratic institutions in the late 1970s, city planners had inherited a city that lagged in nearly every aspect of urban infrastructure. While urban greening was not high on the political agenda, planners in this early democratic period (1978–1980) should be credited for having bought 126 ha of land that would eventually be converted into green spaces, parks and public spaces (Fabre and Huertas 1989). The greening efforts in the first two decades of the democratic government were part of a broader effort to recover and improve public spaces, and address historic inequities in underserved communities.

Planning for green spaces did not become a visible priority until the city began preparations for the 1992 Olympic games. Under the leadership of Mayor Pasqual Maragall, the city aimed to leverage the event to improve the city for residents, even long after the games were over. While this may sound like common sense, this was not how previous Olympic cities had prepared for the event, in which large stadiums or infrastructure were distant or disconnected from the host city. Barcelona's approach emphasised the integration of infrastructure, housing and transportation investments, and this integrated approach was an important legacy of the Barcelona planning model developed during the olimpic period. While Barcelona's urban olimpic transformation has been glorified and sold internationally as a success (Monclús 2003), it is also true that many residents were pushed out via expropriation and demolition to make room for the new Olympic infrastructure (Montaner et al. 2012).

In this section I have provided an overview of the history of urban development in Barcelona, with particular attention to the planning for and creation of green spaces. A rapid review of Barcelona's history helps explain why there are limited green spaces in the city. In the next section, I turn to look at the green spaces that have survived development pressures or have been reclaimed by city residents.

3.4 Historical Typology of Barcelona's Green Spaces

By taking a closer look at the current green space network in Barcelona, I propose a historically informed typology with the aim of creating a theoretically coherent and structured way to think about green spaces in the city. What I propose is only one way to organise and understand these spaces. It is a conceptual framing that is not meant to substitute other types of categorisation such as those based on protection status, function, land use, or jurisdiction. Those other forms of categorisation certainly have their role to play. But it seems worthwhile to consider the origins of these green spaces to help us understand the development trajectory of the city. A historically informed typology may also highlight how Barcelona's green spaces differs from other cities.

My proposed typology categorises urban green spaces as (1) residual spaces, (2) militarised spaces, (3) private gardens (4) industrial spaces, (5) plazas, (6) blue spaces and (7) glamour green. The mixed and multiple uses of sites over the centuries can occasionally muddle boundaries between categories. For example, I would consider the *Turó de la Rovira* as a residual space because its physical condition (steep, rocky and distant) explains why it was hardly inhabited, even though it hosted military installations for a brief period during the Spanish Civil War making it also a military space. Nevertheless, despite the occasionally fuzzy boundaries, thinking about the origin of contemporary urban green spaces is useful because it illustrates the process that created them, and in the case of Barcelona at least, one can see that most green spaces are re-invented spaces. The planned and manicured green spaces one finds at the Turó Park designed by the landscape architect, Nicolas Rubio-i-Tudurí are the exception rather than the rule. Barcelona has no master planned greenspace – no Central Park (New York), Stanley Park (Vancouver), or Chapultepec (Mexico City). In this respect, Barcelona's categorisation of green spaces is notable for what is absent. Most green spaces are there because there was something else first, or there is a good reason why the site was undeveloped in the first place and later emerged as a new configuration to meet new needs.

3.4.1 Residual Spaces

Today, many of Barcelona's parks and green spaces are located on residual sites; on the fringe, spaces in between, or otherwise previously ignored. These sites were steep, rocky and far from the city centre, making them undesirable for urbanization. The mountain and valleys of Collserola fall into this category, although its distance to the city centre makes the Collserola more of a destination site rather than an integrated urban green space. In addition to Collserola, this category would also include the network of hills in Barcelona: *Turó de la Rovira*, *el Carmel*, *Creueta del Coll*, *el Putxet* and *la Piera*. They are islands of green, surrounded by intense compact urban development. These hills were initially too steep and far for development, although with the growth of the city, they eventually became incorporated into the city. One could argue that the mountain of Montjuïc might also be included here, although this space has several centuries of military history which merits its own category.

Of the residual urban green spaces, the *Turó de la Rovira* deserves special attention because the controversy surrounding its current management perfectly encapsulates Barcelona's planning paradox: a well-designed park improvement and restoration (2009–2011) ends up revealing a city secret to a wider public and opening the site to a flood of visitors and partying youth that have made locals feel unwelcome and disenfranchised. Later I will describe in greater detail how new and redesigned green spaces are being appropriated by visitors. For now, I merely want to emphasise that historically residual urban spaces are vulnerable to being transformed into poles of attraction for capital and non-city residents.

3.4.2 Militarised Spaces

Within Barcelona's urban fabric, the two largest green spaces can be found on the mountain of Montjuïc and the Ciutadella Park. Both sites have a military history, and both are closely associated with political repression, imprisonment and control. These were spaces used to house soldiers who pointed cannons at the city, ready to use force to suppress unrest. Both spaces were also reclaimed as part of urban renovation associated with international world exhibitions in 1888 and 1929 respectively. Manuel Solà de Morales describes the symmetry between Montjuïc and the Ciutadella as Barcelona's two ears, providing a delicate balance for the city (Solà-Morales 2008).

In 1873, the Citadel becomes Barcelona's first public park, that is, the *Ciutadella* (City of Barcelona 2013a). Consider what it must have been like to witness the conversion of an odious military barracks into a new public garden – indeed, the first city garden. Writings from the Catalan artist Santiago Rusiñol provide a glimpse into what it must have been like to observe such an unthinkable change:

> In that neighbourhood of *La Ribera* there had been such upheaval that nothing of what had been there was left... As for the citadel, they had razed it to the ground and they had been right to do so. First the walls came down, then the glacis were flattened, later they marked out lines and at last they planted flowers there, and as the flowers came the soldiers left, and the more shade the trees made the more barracks were knocked down, until only two and a half were left, disguised as a palace, there under the trees...and what were once barracks smelling of gunpowder and army rations was now a great strewn bouquet of flowers; what were walls, now carpets of grass; what were bastions, parterres; and what was an accursed tower that had heard so many cries of anguish was now a level space, blotched by the sun and full of children playing in the sand... from the *Pla de Palacio* to *Carrer de les Corts*, the same as in other neighbourhoods, everything was being transformed. (Santiago Rusiñol as quoted in Solà-Morales 2008, p 430).

3.4.3 Private Gardens

Unlike other European cities, Barcelona does not have large palaces or royal gardens. Recall that Barcelona was not the centre of political power when ruled by Bourbon dynasty and Castilian royal families. Without large planned green spaces, and in the context of the perversion of Cerdà's plans, the city has inherited pockets of green. Cloisters and private gardens were the only way for urban residents to enjoy green spaces for most of the city's history (Gordi Serrat 2020). Even today, a considerable amount of urban vegetation remains hidden in cloisters or behind walls in private gardens. However several private gardens have opened to the public. Notable examples would include the *Jardins Rubió-i-Lluc* by the old hospital, *Parc del Laberint de l'Horta*, *Villa Cecilia*, *Palau Robert*, *Jardins d'Enric Sagnier*, *Jardins de Can Ferrer*, and *Jardins de la Villa Amelia*, among others.

3.4.4 Industrial Spaces

Many of Barcelona's urban parks and open spaces are located on former industrial sites. City planners were able to reclaim many of these sites through land purchases, especially in the early democratic period (Fabre and Huertas 1989). These land purchases included well known sites including *La Sedeta, la Espanya Industrial, la Maquinista, La Pegaso, Carbones de Nalon* (on Montjuïc), *Fiat Hispania* (Les Corts), *el Parc de les Aigües* (water storage), and *Creueta del Coll* (mining).

3.4.5 Plazas

Barcelona is well known for its hardscaped squares and plazas that are multifunctional for cultural events and activities but often devoid of vegetation other than the occasional tree (Rowe 2006). In many land use maps, these hardscaped plazas are categorised as an urban green space although one might question the environmental services provided, and in some cases, there may not be any vegetation at all. Some neighbourhoods, such as *Gràcia*, incorporated public squares in their original neighbourhood design, but in many segments of the Old City, these public spaces have been re-claimed from other uses. For instance, the square of *Vicenç Martorell* in the Old City, was cleared by heavy bombing during the Civil War, and with reconstruction too costly, a new public space emerged. The *Plaça Nova* square in front of the Cathedral was produced 1940 for the same reason. Therefore demolition explains many open spaces currently found in the Old City. The open spaces at *Allada Vermell, Forat de la Vergonya, Rambla del Raval* are all sites that were previously built that have been reclaimed as public spaces, often due to pressure from neighbourhood groups, and with varying degrees of vegetation.

3.4.6 Blue Spaces

As a coastal city, Barcelona has 12.7 km of coastline that for much of the twentieth century remained hidden and inaccessible due to rail lines that cut off pedestrian and vehicle access to the coast (Busquets 2005). A major legacy of the 1992 Olympic Games was Barcelona's recovery of its waterfront (Marshall 2004), via the removal of the rail lines which sparked the redevelopment of the waterfront in the Poblenou neighbourhood (Akaltin et al. 2019). The location and design of these spaces along the water merit a special categorization. The *Besós* river corridor would be another blue space recovered for residents with a distinct features and historical processes closely linked to its relationship with water (Barcelona Regional 2018).

3.4.7 Glamour Green

The new millennium marked a turning point in the evolution of Barcelona's green space planning. As Barcelona consolidated its place as a new global city, pressure mounted to design for the international market in which parks and green spaces become desirable amenities that capture the attention of the global elite and help lubricate the sale of luxury apartments. This shift is most evident in the Forum-Diagonal Mar development (2004), where the neighbourhood plan explicitly rejected Cerdà's grid and included a North American-style shopping mall. The launching of Diagonal Park opened a new era in green space planning for the city: one that caters to international interests and private investors, rather than mitigating historic inequities. Attracting star architects like Jean Nouvel (Parc del Poblenou) seems to confirm suspicions that the city aims to leverage new green space designs for a global audience, presumably to attract high skilled workers in the knowledge economy, rather than serving the needs of existing residents (Pérez del Pulgar et al. 2020). The proposed street greening of the Cristobal de Moura corridor that traverses the remaining undeveloped industrial segment of Poblenou appears to be an overt strategy to use public funds to attract private redevelopment capital. In projects such as these, neighbours are raising legitimate questions about who will capture the increase in land values associated with these greening projects (Kotsila et al. 2020). While not all recent greening efforts have international ambitions, 'glamour greening' has generated a deep suspicion among residents. Debates about the extent to which city greening projects will benefit or exclude residents are now at the centre of the discussion. Later in this chapter I review how these debates are influencing current and future proposals for city greening.

3.5 Greening for the Resilient City

Having provided this historical context, I now shift attention to the most recent greening efforts developed by city officials. In this section I discuss the contemporary debates on city greening in Barcelona, especially as they relate to the new Superblock Barcelona proposal and the associated critique of urban greening. I also reflect on how our conceptualization and management of greenspaces might change as a result of the pandemic.

3.5.1 Current Plans and Strategies

The urban history described above makes it clear why Barcelona has inherited a green space network that is small, fragmented and disconnected. Most parks do not have management plans, and the city recognises that 57% of its greenspaces are small (<1500 m^2) (City of Barcelona 2013a). Cerdà's beloved *Eixample* grid today falls short on most measures of urban greening. Over one third (33%) of the surface area in the *Eixample* are devoted to road space while only 7% is devoted to urban parks, producing an unbalanced ratio of 4.8 times more road space than green space. The entire city of Barcelona fares only a bit better, with 13% of the city categorised as urban parks and 21% as roads (City of Barcelona 2020a). City planners now aim to weave together the historical patchwork of green spaces combined with the use of 'greening acupuncture' with green walls, green buildings, rooftop terraces, street re-designs. Inserting vegetation and greenspaces into the city will moderate the urban heat island effect, reduce energy demands, improve thermal comfort for residents, attract biodiversity, improve water management, and create spaces that contribute to our physical and mental health. For all these reasons, urban greening is a central strategy for creating a climate resilient city.

The early policy documents related to climate resilience planning pay only minor attention to green spaces (City of Barcelona 2015). For example, the policy measure on resilience does not discuss land use or green space issues, but instead focuses is on its capacity to respond to shocks (City of Barcelona 2016). However it is clear that city planners and architects appreciate the need to increase green space as part of its climate resilient strategy and the incorporation of green spaces figures more prominently in the 2020 Climate Emergency Declaration (City of Barcelona 2020b), and the 2018 Climate Plan (City of Barcelona 2018). The creation of more green spaces in the city is one of the 18 strategic actions set out in the 2018–2030 Climate Plan. The City sets out the goal of obtaining 160 ha of new green space by 2030, which would provide at least 1 m^2 additional per inhabitant (City of Barcelona 2018). Furthermore, the *Trees for Life: Master Plan for Barcelona's Trees* establishes several goals such as (1) increasing the city's tree cover by 5%, so that 30% of the city's surface area is covered by trees; (2) ensuring that 40% of tree species are adapted to climate change; and (3) achieving a biodiverse tree heritage in which no single species accounts for more than 15% of the total population within the urban area (City of Barcelona 2017). While these targets are admirable, they do not anticipate meeting their goal until 2037 and there is certainly room to be more ambitious. However the city is currently revising its biodiversity plan *Pla Natura 2021–2030* in which there appear to be more ambitious urban greening objectives, particularly in Cerdà's Eixample district.

Nevertheless, there is perhaps too much importance given to the metric 'green space per inhabitant'. This metric is too abstract, does not address the issue of everyday accessibility and hides neighbourhood differences. Instead, it might be more useful to use the 3-30-300 rule of thumb for urban greening that proposes that every city resident should be able to see at least 3 trees from their home, have 30% canopy cover in every neighbourhood, and live 300 m from an urban park or greenspace (Konijnendijk van den Bosch 2021). This rule of thumb works at different scales, and is based on research that shows the health and wellbeing benefits of urban trees and vegetation (van den Bosch and Nieuwenhuijsen 2017).

A central piece of the city's future urban greening strategy is based on the *Superblock Barcelona* proposal (City of Barcelona 2021). In March 2021, Barcelona

Mayor Ada Colau and her team announced the winners of the Superblock design competition for the reconfiguration of four streets in the *Eixample* grid: *Consell de Cent, Compte Borrell, Rocafort* and *Girona* – as well as the conversion of four vehicular intersection into green plazas. The project will redesign the streets to become pedestrian friendly, healthier and greener. The protagonist of this transformation is *Consell de Cent*, a street 2.8 km in length aiming to be the pioneer green corridor, and the test bed of ideas for a much wider transformation. With a budget of 37.8 million Euros, the immediate goal is to complete the green corridor on *Consell de Cent* and the four plazas by February 2023.

And while this proposal is ambitious, the vision goes beyond this particular project. As illustrated by the name of the project, Superblock Barcelona does not intend to be a one-off city greening project, rather they are talking about a new urban model for Barcelona's *Eixample* district. The idea is to scale up the existing Superblock experiences that have been tested in the Poblenou and Sant Antoni neighbourhoods which consisted of street calming and greening for nine blocks of Cerdà's grid (Rueda 2019; Zografos et al. 2020, see Chap. 4 for the Poblenou superblock project). In Superblock Barcelona, city planners would like to apply the same principles to the entire *Eixample* district to distribute the same pedestrian-oriented experience to the entire neighbourhood, and therefore most of the city. They are talking about a full overhaul of what it might feel like to be in the historic grid designed by Cerdà.

3.5.2 Greening, Tourism and Gentrification

Within 24 hours of Mayor Colau announcing the ambitious plans for Superblock Barcelona, voices on social media already began to question the project as an exercise in greenwashing. Is this project a manoeuvre to deflect criticisms that the city remains in persistent incompliance with European air quality standards? Is this project a strategy to capture financial investment or brand the city as green? Who will benefit and who will be excluded? In this city, few urban greening projects can avoid questions of distributional justice and gentrification (Kostilla et al. 2021).

Perhaps unlike anywhere else, concerns about greenwashing and green gentrification are front and centre in Barcelona. The research team at the Barcelona Lab for Urban and Environmental Justice (www.bcnuej.org) has certainly helped focus attention on the relationship between city greening, gentrification and exclusion (Anguelovski et al. 2020; Oscilowicz et al. 2021). In my view, it would still be premature to dismiss Superblock Barcelona as an effort in greenwashing. Of course, there are good reasons why citizens are concerned about the local impacts of urban greening. While these beautification projects strive to provide locals with green amenities, they may also produce unexpected and unintended consequences.

Since the late 1990s, Barcelona has emerged as a global city – a hub on the Mediterranean, a city of design, finance, innovation, culture and delight. The city's renaissance is closely linked to bold urban transformations often referred to as the

'Barcelona Model' of urban planning and design (Marshall 2004; Rowe 2006; Montaner et al. 2014). And yet despite Barcelona's planning success—or perhaps because of it—there is growing discontent about how the city is being transformed (Montaner et al. 2012). While many residents and visitors recognise the city improvements over the last 30 years, there is unease about the market logic that is driving these changes and concern that locals are not benefiting (Marshall 2004; Delgado 2007). Barcelona's success has attracted millions of visitors each year, reaching 12 million visitors in 2019 (Minder 2020). Many are worried that the city's carrying capacity for tourists has been reached (López 2017; Milano et al. 2019). The city centre is being transformed to meet the needs of the visitors, and tourism is displacing everyday life and everyday residents (Arias-Sans and Quaglieri Domínguez 2016; Oscilowicz et al. 2020).

This puts forth a paradox: Barcelona's urban success has surely made it a more enjoyable place to visit, but perhaps a less enjoyable place to live, as neighbourhood space is converted to lucrative commercial uses that target visitors, rather than residents, and green spaces are filled with a constant flow of influencers and instagrammers. Residents fear that Barcelona is being transformed into a theme park, catering only to temporary travellers. 'Barcelona is over', writes Sergi Picazo, 'at least the Barcelona that we knew, is over. It is no longer a city: it is a product' (Picazo 2017). Frustration at foreign visitors motivated activists to take direct action. In 2017, anti-tourism activists attacked a City Bus Tour filled with curious tourists who sat watching while the masked vandals slashed tires and spray painted 'Tourism kills neighborhoods' (Burgen 2017; Hughes 2018).

The public space re-design of *Turó de la Rovira* (2009–2011) perhaps most vividly illustrates the appropriation of successful green space planning by global tourism. *Turó* translates as *hill* in Catalan, and during the Spanish Civil War, the Republican army used the elevated site to install anti-aircraft gun emplacements to defend the city from bombings by the fascist Italian airforce. The site was abandoned, and the steep hills became the site of informal housing. To recover city heritage, memory and landscape, the City History Museum of Barcelona (MUHBA) led a site restoration project that invested in connecting the green space to surrounding hills and simultaneously respected the wild and isolated experience on the hill. New wayfinding and interpretative signs inform visitors about the informal housing that lived on the steep slopes only a few decades ago. The design team led by Imma Jansana and Jordi Romero did an excellent job creating a unique, educational and memorable public space, and the project was awarded the European Public Space prize in 2012 (Gray 2015).

When Lonely Planet included *Turó de la Rovira* in its guidebook as one of the best places to get a birds-eye view of the city, the site was overwhelmed with visitors. The secret was out, and tourists appropriated the space as theirs to picnic, party and play loud music, leaving local residents disenfranchised. If you hike up to the *Turó de la Rovira* today, you will be greeted by graffiti informing you that 'Tourism Kills' or 'Tourists Go Home', yet this has hardly dissuaded travellers who are determined to get their perfect selfie with the city. The case of *Turó de la Rovira* illustrates how attractive urban design and park restoration may attract visitors who appropriate the space, and local interests may be eclipsed when tourists flock to a landscape that is glamorised in social media.

In response to growing discontent over 'touristification', the progressive City government declared a moratorium on the construction of new hotels in the city centre, revised by-laws that regulated outdoor seating by bars and restaurants and made a series of promises oriented at planning for residents – not visitors. Greening the city is now part of the resident-focused agenda to make the city healthier and more liveable and adapted to climatic challenges (City of Barcelona 2013b). Yet the renewed interest in city greening is also viewed with scepticism (Anguelovski et al. 2018). To what extent are investments in city-greening and beautification driven by part of a larger neoliberal project? Who benefits most from the greening agenda in Barcelona, and have past greening efforts exacerbated environmental inequities or mitigated them? These are the outstanding questions that need to be answered to transition to a greener and more resilient city.

3.5.3 Urban Greening in the Post-pandemic City

Looking ahead, the COVID-19 pandemic may re-calibrate our expectations about the role of urban green spaces in cities, especially with regard to how much, how close and how crowded they should be (Honey-Rosés et al. 2020; Venter et al. 2020). The pandemic has forced us to scrutinise the relationship between urban density, green spaces, and health. Extensive research has already identified the health benefits of green spaces (Ward Thompson et al. 2012; Bratman et al. 2015). Now more attention might be given to who is using these spaces (Pérez del Pulgar et al. 2020), issues of carrying capacity, and the impact of high-use on perceptions and wellbeing (Honey-Rosés and Zapata 2021). Post-pandemic, urban residents might have greater expectations about the accessibility of green spaces from their home and the ability to find places for refuge and calm. These rising expectations about how much and how accessible green spaces should be in cities is coinciding with a greater sense of urgency for action on climate change. Cities that successfully leveraged the pandemic to incorporate more greenspaces into cities will emerge as a more liveable, healthier and climate resilient city (Honey-Rosés et al. 2020).

3.6 Conclusion

The chapter has traced the history and evolution of urban green space planning in Barcelona to contextualise current discussions about urban climate resilience. Advocates for resilient cities often call for a profound re-organisation of our urban landscape. Given this ambitious goal, it is worth considering how our proposals fit along a continuum of ideas and proposals that have been in circulation since at least the late nineteenth century, if not earlier.

For both scholars and practitioners, a historical analysis helps us distinguish between truly new and innovative ideas from old ones. Which proposals push us in new directions, and which are merely reconfigurations of old material in new boxes? We should not shun ideas because they have been recycled, rather we should merely appreciate that they are part of an ongoing dialogue about the cities we want. As argued by scholars aiming to historicise the discussion on climate resilience (Carey and Garone 2014; Adamson et al. 2018), a historical appreciation will also give us the perspective, humility and motivation to work for a better urban future.

Barcelona residents have lived for centuries with limited access to green spaces and parks. This historic deficit in green space has motivated planners and activists to petition for improved access to open space. Most green spaces in the city today are residual spaces, previously deemed unsuitable for human development, or reclaimed industrial spaces or private gardens that have been opened to the public. Starting around 2000, the city welcomed a new typology of greenspace, glamour green, produced by a capitalist logic targeting high-end luxury homeowners. The appropriation of these spaces by temporary visitors and tourists makes Barcelona unique in discussions on the green gentrification and exclusion.

Much has been said about the transition to resilience, but a historical analysis forces us to ask about our starting point. Discussion on resilience tend to be future oriented, where cities need to go, and what needs to be done. It is also useful to think about where a particular city has been because its capacity to evolve and the direction that it may travel is contingent on historic constraints.

Urban resilience has become the new catch phrase – giving planners and the public the sense that our plans and projects are preparing for our uncertain future. Urban resilience will help rally new followers to old agendas. And while the term is difficult to define, there is agreement that it is useful. Similarly, there seems to be consensus that a greener, more naturalised city will be better prepared for the climate-related shocks of the future.

Acknowledgements This chapter benefited from extraordinary feedback and suggestions from Isabel Ruiz-Mallén, Mar Satorras and Zaida Muxi on earlier drafts of the manuscript. The author was supported by the Ramón y Cajal Fellowship (Ministerio de Ciencia y Universidades RyC-2019-027279-I) and this work contributes to the ICTA-UAB "María de Maeztu" Programme for Units of Excellence of the Spanish Ministry of Science and Innovation (CEX2019-000940-M). All errors are responsibility of the author.

References

Adamson GCD, Hannaford MJ, Rohland EJ (2018) Re-thinking the present: the role of a historical focus in climate change adaptation research. Glob Environ Chang 48:195–205. https://doi.org/10.1016/J.GLOENVCHA.2017.12.003

Akaltin D, Chen J, Cho W et al (2019) Life in Poblenou: observing spaces in transition. Vancouver. doi:10.14288/1.0384917. URL: http://hdl.handle.net/2429/72184.

Anguelovski I, Connolly JJT, Masip L, Pearsall H (2018) Assessing green gentrification in historically disenfranchised neighborhoods: a longitudinal and spatial analysis of Barcelona. Urban Geogr 39:458–491. https://doi.org/10.1080/02723638.2017.1349987

Anguelovski I, Brand AL, Connolly JJT et al (2020) Expanding the boundaries of justice in urban greening scholarship: toward an emancipatory, antisubordination, intersectional, and relational approach. Ann Am Assoc Geogr 0:1–27. https://doi.org/10.1080/24694452.2020.1740579

Arias-Sans A, Quaglieri Domínguez A (2016) Unravelling airbnb: urban perspectives from Barcelona. In: Paolo Russo, A & G. Richards (Eds) Reinventing the local in tourism: producing, consuming and negotiating place, Chanel View Publications, Bistol p 209–228. IBSN 9781845415709. https://doi.org/10.21832/9781845415709-015

Barcelona Regional (2018) Agenda Besòs. Barcelona

Bratman GN, Daily GC, Levy BJ, Gross JJ (2015) The benefits of nature experience: improved affect and cognition. Landsc Urban Plan 138:41–50. https://doi.org/10.1016/j.landurbplan.2015.02.005

Burgen S (2017) Barcelona anti-tourism activists vandalise bikes and bus. Guard

Busquets J (2005) Barcelona. The urban evolution of a compact city. Harvard University. Graduate School of Design, Cambridge

Carey M, Garone P (2014) Forum introduction. Environ Hist Durh N C 19:282–364. https://doi.org/10.1093/envhis/emu004

City of Barcelona (2013a) Greenspace and biodiversity plan 2020 (original title: Pla del verd i de la biodiversitat de Barcelona 2020). Barcelona, Catalonia

City of Barcelona (2013b) Barcelona green infrastructure and biodiversity plan 2020 (Original title: plan del verd i de la biodiversitat de Barcelona 2020). Barcelona

City of Barcelona (2015) Compromís de Barcelona pel Clima. Barcelona + Sostenible

City of Barcelona (2016) Resiliència urbana: mesura de govern

City of Barcelona (2017) Trees for life: master plan for Barcelona's trees 2017–2037. Barcelona

City of Barcelona (2018) Pla Clima 2018–2030. Barcelona

City of Barcelona (2020a) Anuari estadístic de la ciutat de. Barcelona

City of Barcelona (2020b) Climate emergency declaration. Barcelona

City of Barcelona (2021) Superblock Barcelona: towards the city we want. Barcelona

Delgado M (2007) La ciudad mentirosa: fraude y miseria del "modelo Barcelona". Catarata, Barcelona

Fabre J, Huertas JM (1989) Barcelona, la construcció d'una ciutat, 1888–1988. Editorial Plaza & Janés, Barcelona

Gills B, Morgan J (2020) Global climate emergency: after COP24, climate science, urgency, and the threat to humanity. Globalizations 17:885–902. https://doi.org/10.1080/14747731.2019.1669915

Gordi Serrat J (2020) El paper dels arbres en l'evolució urbana de Barcelona (1050–1992). Treballs la Soc Catalana Geogr:9–38. https://doi.org/10.2436/20.3002.01.198

Gorostiza S, Martí Escayol MA, Barriendos M (2021) Controlling water infrastructure and codifying water knowledge: institutional responses to severe drought in Barcelona (1620–1650). Clim Past 17:913–927. https://doi.org/10.5194/cp-17-913-2021

Grau-Satorras M, Otero I, Gómez-Baggethun E, Reyes-García V (2016) Long-term community responses to droughts in the early modern period: the case study of Terrassa, Spain. Ecol Soc 21. https://doi.org/10.5751/ES-08232-210233

Gray, D (2015) Europe City: Lessons from the European Prize forUrban Public Space. CCCB, Barcelona (199 pgs)

Honey-Rosés J, Zapata O (2021) The impact of residential densification on perceptions of public space. J Am Plan Assoc 87(2):282–295. https://doi.org/10.1080/01944363.2020.1846597

Honey-Rosés J, Anguelovski I, Chireh VK et al (2021) The impact of COVID-19 on public space: an early review of the emerging questions – design, perceptions and inequities. Cities Heal 00:1–17. https://doi.org/10.1080/23748834.2020.1780074

Hughes R (1992) Barcelona. Alfred A. Knopf, New York

Hughes N (2018) 'Tourists go home': anti-tourism industry protest in Barcelona. Soc Mov Stud 17:471–477. https://doi.org/10.1080/14742837.2018.1468244
Johnson S (2006) The ghost map: a street, a city, an epidemic and the hidden power of urban networks. Penguin Books, London
Kallis G (2011) In defence of degrowth. Ecol Econ 70:873–880. https://doi.org/10.1016/j.ecolecon.2010.12.007
Konijnendijk van den Bosch C (2021) Promoting health and wellbeing through urban forests – introducing the 3-30-300 rule. In: IUCN. https://iucnurbanalliance.org/promoting-health-and-wellbeing-through-urban-forests-introducing-the-3-30-300-rule/
Kostilla P, Oscilowicz E, Segulova F et al (2021) Barcelona's greening paradox as an emerging global city and tourism destination. In: Anguelovski I, Connolly JJT (eds) The green city and social injustice: 21 Tales from North America and Europe. Routledge, Abingdon/New York
Kotsila P, Anguelovski I, Baró F et al (2020) Nature-based solutions as discursive tools and contested practices in urban nature's neoliberalisation processes. Environ Plan E Nat Sp 0:251484862090143. https://doi.org/10.1177/2514848620901437
Lewis SL, Maslin MA (2015) Defining the Anthropocene. Nature 519:171–180. https://doi.org/10.1038/nature14258
López H (2017) El Gòtic no quiere ser Venecia, y el Poblenou no quiere ser el Gòtic. El Periódico
Magrinyà F, Marzá F (2017) Cerdà. 150 years of modernity. Actar Publishers, Barcelona
Marshall T (2004) Transforming Barcelona. Routledge, London
Martí F, Moreno E (1974) Barcelona ¿a dónde vas? Editorial Dirosa, Barcelona
Milano C, Novelli M, Cheer JM (2019) Overtourism and degrowth: a social movements perspective. J Sustain Tour 27:1857–1875. https://doi.org/10.1080/09669582.2019.1650054
Minder R (2020) Wandering a grand hotel emptied by coronavirus, and checking 1,400 taps. New York Times. p 1–5
Monclús F-J (2003) The Barcelona model: and an original formula? From 'reconstruction' to strategic urban projects (1979–2004). Plan Perspect 18:399–421. https://doi.org/10.1080/0266543032000117514
Montaner JM, Álvarez F, Muxí Z (2012) Arxiu crític model Barcelona. Ajuntament de Barcelona, Barcelona
Montaner JM, Álvarez F, Muxí Z, Casanovas R (2014) Reader modelo Barcelona 1973–2013. Comanegra, Barcelona
Neumann MB, Rieckermann J, Hug T, Gujer W (2015) Adaptation in hindsight: dynamics and drivers shaping urban wastewater systems. J Environ Manag 151:404–415. https://doi.org/10.1016/j.jenvman.2014.12.047
Oscilowicz E, Honey-Rosés J, Anguelovski I et al (2020) Young families and children in gentrifying neighbourhoods: how gentrification reshapes use and perception of green play spaces. Local Environ 25:765–786. https://doi.org/10.1080/13549839.2020.1835849
Oscilowicz E, Lewartowska E, Levitch A et al (2021) Policy and planning tools for urban green justice: fighting displacement and gentrification and improving accessibility and inclusiveness to green amenities. Barcelona, Spain
Pérez del Pulgar C, Anguelovski I, Connolly J (2020) Toward a green and playful city: understanding the social and political production of children's relational wellbeing in Barcelona. Cities 96:102438. https://doi.org/10.1016/j.cities.2019.102438
Picazo S (2017) McDonald' s i AirBnB al Carmel. El Crític
Rees W, Wackernagel M (1996) Urban ecological footprints: why cities cannot be sustainable and why they are key to sustainability. Environ Impact Assess Rev 9255:223–248
Rockström J, Steffen W, Noone K et al (2009) A safe operating space for humanity. Nature 461:472–475. https://doi.org/10.1038/461472a
Rowe PG (2006) Building Barcelona: a second Renaixença. Barcelona Regional, Barcelona
Rueda S (2019) Superblocks for the design of new cities and renovation of existing ones: Barcelona's case. In: Nieuwenhuijsen M, Khreis H (eds) Integrating human health into urban and transport planning. Springer, Cham, pp 135–153

Satorras M, Ruiz-Mallén I, Monterde A, March H (2020) Co-production of urban climate planning: insights from the Barcelona climate plan. Cities 106:102887. https://doi.org/10.1016/j.cities.2020.102887
Satterthwaite D, Dodman D (2013) Towards resilience and transformation for cities within a finite planet. Environ Urban 25:291–298. https://doi.org/10.1177/0956247813501421
Sennett R (2018) Building and dwelling: ethics for the city. Penguin Random House, London
Servigne P, Stevens R (2020) How everything can collapse. Policy Press, Cambridge
Solà-Morales M (2008) Ten lessons on Barcelona, 2nd edn. Col.legi d'Arquitectes de Catalunya, Barcelona
van den Bosch M, Nieuwenhuijsen M (2017) No time to lose – green the cities now. Environ Int 99:343–350. https://doi.org/10.1016/j.envint.2016.11.025
Venter Z, Barton D, Gundersen V et al (2020) Urban nature in a time of crisis: recreational use of green space increases during the COVID-19 outbreak in Oslo, Norway. Environ Res Lett 15. https://doi.org/10.1088/1748-9326/abb396
Ward Thompson C, Roe J, Aspinall P et al (2012) More green space is linked to less stress in deprived communities: evidence from salivary cortisol patterns. Landsc Urban Plan 105:221–229. https://doi.org/10.1016/j.landurbplan.2011.12.015
Wise RM, Fazey I, Stafford Smith M et al (2014) Reconceptualising adaptation to climate change as part of pathways of change and response. Glob Environ Chang 28:325–336. https://doi.org/10.1016/J.GLOENVCHA.2013.12.002
Ynfante J (1974) Los negocios de Porcioles. Las Sagradas Familias de Barcelona. Ediciones Monopolio, Tolouse
Zografos C, Klause KA, Connolly JJT, Anguelovski I (2020) The everyday politics of urban transformational adaptation: struggles for authority and the Barcelona superblock project. Cities 99:102613. https://doi.org/10.1016/j.cities.2020.102613

Chapter 4
Urban Transformational Adaptation: Contestation and Struggles for Authority in the Pilot Barcelona Superblock of Poblenou

Isabelle Anguelovski, Christos Zografos, Kai A. Klause, and James J. T. Connolly

Abstract Cities' increasing vulnerability to the effects of climate change calls for greater attention to urban transformational adaptation as a path towards environmental and social resilience. Through the initial Superblock project in the Poblenou neighbourhood of Barcelona – whereby traffic pacification is combined with new open space and transit network improvements – we analyse municipal efforts aligning transformational land use planning with climate adaptation objectives and the pushback again them. The initial opposition to the project embodied the everyday political struggle for municipal authority, that is, clashes over visions for the future Barcelona and who can define them and own them. Urban transformation – and the contestation around it – is thus not only about negotiating environmental and quality-of-life benefits that are ostensibly the target of interventions, but also

I. Anguelovski (✉)
ICREA, Barcelona Lab for Urban Environmental Justice and Sustainability, Institute of Environmental Science and Technology (ICTA), Universitat Autònoma de Barcelona, Bellaterra, Barcelona, Spain
e-mail: Isabelle.Anguelovski@uab.cat

C. Zografos
Department of Political and Social Sciences, Universitat Pompeu Fabra, Barcelona, Spain

K. A. Klause
Barcelona Lab for Urban Environmental Justice and Sustainability, Institute of Environmental Science and Technology (ICTA), Universitat Autònoma de Barcelona, Bellaterra, Barcelona, Spain

J. J. T. Connolly
Barcelona Lab for Urban Environmental Justice and Sustainability, Institute of Environmental Science and Technology (ICTA), Universitat Autònoma de Barcelona, Bellaterra, Barcelona, Spain

School of Community and Regional Planning, University of British Columbia, Vancouver, BC, Canada

I. Ruiz-Mallén et al. (eds.), *Urban Resilience to the Climate Emergency*, The Urban Book Series, https://doi.org/10.1007/978-3-031-07301-4_4

competitive urbanism and related short-term political gains. Of particular importance is how civic and political contestation over the authority of *climate champions* can jeopardise not only transformational adaptation achievements, but also the political survival of those champions. Transformational adaptation can be slowed down not only by the fear for the material and political effects of transformation per se, but also because of how key stakeholders and the residents around them contest who has the authority to decide for *the common good*. This chapter is a modified and updated version of the Zografos et al (Cities 99:102613. Copyright (2020), with permission from Elsevier, 2020) article previously published in Cities, Elsevier.

Keywords Transformational adaptation · Competitive urbanism · Mobility · Public space · Superblocks

4.1 Introduction

Although cities are increasing the resources dedicated to addressing the risks and impacts of climate change, a worrying disconnection exists between the prescriptions of climate science and the reality of deeply rooted, unsustainable urban development patterns. Climate scientists and policy analysts assert that, because cities are high producers of greenhouse gas emissions and urban residents are under high risk of negative climate impacts (IPCC 2014; Revi et al. 2014), there are strong untapped opportunities for municipal climate adaptation strategies to be bold and make rapid progress (Araos et al. 2016).[1] Indeed, ongoing initiatives for urban climate adaptation motivated by increasing experiences with natural hazards as well as cities' desire to show leadership in climate action and connect with international organisations and networks implies that the conditions for such progress are present (Carmin et al. 2013).

In contrast to incremental approaches to climate change adaptation, transformational urban adaptation seeks to respond to climate change by pursuing fundamental and structural shifts in urbanisation processes and outcomes, which are meant to challenge unsustainable development pathways in a radical way. Although advocacy for transformational approaches is gaining relevance, the understanding of what transformational adaptation should look like varies (Mustelin and Handmer 2013; IPCC 2014; Lonsdale et al. 2015). Keeping in mind that there are multiple interpretations (Mustelin and Handmer 2013; Lonsdale et al. 2015), certain core characteristics can be attributed to transformational urban adaptation. First, transformational actions involve non-linear alterations at an enlarged scale or

[1]Araos et al. (2016) illustrate that only 15% of the 401 municipalities with more than 1 million inhabitants report adaptation initiatives.

intensity that reorder and/or relocate systems, transform places, and shift locations (Kates et al. 2012). Second, they address *underlying failures of development, including increasing greenhouse gas emissions by linking adaptation [and] mitigation* (IPCC 2014 quoted in Pelling et al. 2014: 2). Third, they seek fundamental alterations *within* a system (like a city) that itself produces climate change vulnerability both elsewhere and within its contours. In doing so, they seek to affect both local adaptive capacity and global mitigation. Fourth, they confront generative causes of vulnerability to climate change by engaging with the politics of managing risk (Ribot 2011; Connolly 2018). Perhaps unsurprisingly, there are few concrete examples of transformational adaptation at the city scale, and non-system-affirming interventions (Lubell et al. 2009; Lang and Rothenberg 2017). Indeed, finding cases of transformational adaptation is difficult in general (O'Brien et al. 2015), and in particular, it is difficult to find cases of transformational initiatives related to climate change where the promoters of those initiatives explicitly frame them as *transformational adaptation.*

Expectedly, the path to transformation is not smooth. The climate change adaptation literature[2] identifies several barriers to transformation: cognitive limitations (Park et al. 2012; Pelling et al. 2015; Lonsdale et al. 2015); uncertainties about the level of benefits from transformation (Kates et al. 2012) as well as transaction and opportunity costs, and the costs of risks and unexpected side-effects of transformations (Kates et al. 2012; Rickards and Howden 2012); path dependency (Hadarits et al. 2017; Termeer et al. 2017); the difficulty to transform institutions designed to pursue ends contradicting transformational priorities (Godfrey-Wood and Otto Naess 2016) and reward incremental action (Pelling et al. 2015; Kates et al. 2012); short-termism in responding to climate pressures (Ribot 2011); organisational limitations (Termeer et al. 2017); collective action barriers (Lonsdale et al. 2015); the complexity of transforming multiple spheres of human experience and interacting socio-ecological spheres (O'Brien 2012; Lonsdale et al. 2015; Pelling et al. 2015; Gillard et al. 2016); a lack of attention to institutionalised inequities responsible for vulnerabilities and urban environmental inequities in the first place and to elite and private interest capture of the adaptation and development agenda (Anguelovski et al. 2016); and, the relevance of established values and personal beliefs that can mobilise resistance to transformation (Kates et al. 2012; O'Brien 2012; O'Brien and Sygna 2013; Park et al. 2012; Pelling et al. 2015), such as place attachment and occupation attachment that can be negatively correlated to transformational capacity (Marshall et al. 2012; Park et al. 2012; Hadarits et al. 2017).

In this chapter, we are particularly interested in examining political barriers and power relations, another group of barriers to transformation identified in the literature, but so far underexamined because of the rare, on-the-ground instances of

[2] With the exception of few empirical studies that typically focus on agriculture in the Global North (Hadarits et al. 2017; Marshall et al. 2012; Park et al. 2012) – two exceptions are Termeer et al. (2017) who analyse a more comprehensive flood protection policy initiative in the Netherlands, and Ziervogel et al. (2016) who explore transformative capacity in urban settlements in South Africa), this transformational adaptation literature is mostly conceptual.

transformational urban practice. There is a broad consensus among critical work on transformational adaptation that power, politics and interests can pose daunting barriers to transformation (Pelling and Manuel-Navarrete 2011), that deliberate transformations require *changes to entrenched systems maintained and protected by powerful interests* (O'Brien 2012), and that regardless of the approach taken towards transformation, unequal power relations can prevent transformative change (Godfrey-Wood and Otto Naess 2016). Power distributions in existing institutions combined with *lock-in to unsustainable development pathways* (Pelling et al. 2015) point to the need for confronting systemic barriers in the course of realising transformation (Glavovic 2008). For example, system stagnations, such as vicious circles, exasperating delays, or an escalated conflict, are a case of barriers for changing environmental dynamics that can be rooted in vested interests (Termeer et al. 2017). Power and politics are also expressed through the construction of subjectivity, knowledges, and the operation of authority (Nightingale 2018), phenomena that are reproduced and contested with adaptation practices *in ways that can open up or close down space for transformational adaptation* (Eriksen et al. 2015). This emphasis on the political sphere also draws attention to the crisis of capitalism (O'Brien and Sygna 2013). This includes the capacity of neo-liberalism to create *systemic constraints that incapacitate Western democracies from going beyond incremental adaptation efforts* (Fieldman 2011 quoted in Termeer et al. 2017). Related work also analyses the capacity of the political economy of capitalism to appropriate transformation as an opportunity for capital accumulation (Whitehead 2013; Godfrey-Wood and Otto Naess 2016; Warner and Kuzdas 2017), and the foregoing of transformative adaptation interventions in vulnerable places in favour of incremental ones in ways that allow maintaining rent-seeking projects elsewhere (Zografos 2017).

Indeed, many adaptation programmes are not politically neutral (Shi et al. 2016; Chu 2017), require strategic approaches (Chu et al. 2017), and are designed to work within neoliberal development pathways and practices (Whitehead 2013; Revi et al. 2014; Chu 2017) as a result of which they potentially reproduce uneven socio-economic vulnerability (Eriksen and Brown 2011; Chu 2015; Anguelovski et al. 2016). Efforts to bring about urban transformational adaptation must necessarily confront those dynamics by considering the power relations that underlie urban development pathways, and which (re)produce uneven vulnerability to climate change and environmental injustice (Pelling 2010; Ribot 2011; O'Brien 2012; Bahadur and Tanner 2014; Taylor 2014; Zografos et al. 2016).

If large-scale, systemic, non-linear changes that cross the threshold of transformational adaptation are to take place, as others have mentioned (Bicknell et al. 2009), scholarly attention needs to focus on how the political constraints and institutionalised power relations that shape urbanisation and exacerbate unevenly experienced risk affect climate policy. From this perspective, the urbanisation to be transformed is *an expression of intersecting regimes of social power* (Whitehead 2013: 1348) that needs to be reconfigured to integrate inclusiveness, equity, and resilience into urban climate infrastructure and design processes (Bahadur and Tanner 2014; Revi et al. 2014; EEA Report 2016; Chu 2017). We thus believe that

adaptation, then, is inextricably entangled with the question of who has authority – understood as the legitimate ability to exercise and control power – over land use and beyond. Struggles for authority may open up or close down spaces for transformational adaptation through, for example, competition over the ability to set agendas in the process of urban environmental governance (Sikor and Lund 2009; Fairhead et al. 2012; Nightingale and Ojha 2013; Eriksen et al. 2015). Given contested claims to authority, whether urban climate adaptation becomes transformational or not is partly dependent on the extent to which some visions of the future city become legitimised within city planning and development processes (Eriksen et al. 2015). In that sense, issues related to environmental policy and planning per se might become secondary in the fight for transformational planning.

Although the current scholarship has produced elaborate and highly valuable conceptual explanations of how power relations and struggles, and interdependent causes may affect transformational adaptation, empirical research on barriers to adaptation initiatives of a transformational nature in urban areas can hardly be found. In their review of the literature on transformational adaptation, Lonsdale et al. (2015) also note *a dearth of examples of transformational adaptation that enquire with sufficient depth and honesty into the real, messy practice involved in trying to transform an existing system* (p. 30), and that traditional case study approaches are reluctant *to record aspects that might be considered mistakes or failure, although these parts are where the potential for greatest learning lies* (p. 28). In our views, understanding resistance processes is key to moving transformational adaptation knowledge forward.

This chapter attempts to address those gaps in knowledge, by examining what happens when urban authorities try to implement transformational land use policy but face resistance, as in the case we present in this paper. We ask the questions: What sort of barriers emerge and how, when authority tries to pursue transformational action at the city level and faces resistance? And what do these political dynamics tell us about best ways of implementing transformational adaptation at the urban level? Our analysis responds to these core questions through the analysis of the politics of a conflict around a flagship transformational urban intervention – the superblock project in Barcelona, Spain.

4.2 A Pilot Project in Barcelona: The Poblenou Superblock

Our research examines the early implementation of the Barcelona superblock programme's pilot project in the district of Poblenou (also called SP9) – henceforth, the superblock.[3] While Barcelona's superblocks is a unique case of an early urban intervention that was framed in the language of transformational adaptation and integrated into the city's climate strategy, it is representative of large-scale initiatives

[3] *Superilla* in Catalan.

aiming at addressing climate challenges within other European Mediterranean cities, including those seeking to mitigate heat island effect, coastal erosion, and flooding. Barcelona's superblock programme has been an ambitious project that aims at transforming multiple dimensions of the urban living experience. Although urban greening, sustainable mobility, increased public space use and citizen participation in planning are key programme objectives (see Chap. 8 for the co-production process of the Barcelona Climate Emergency Declaration), the superblock project has become explicitly integrated into the city's Climate Action Plan and Barcelona's Climate Commitment initiative mostly due to its capacity to address the urban heat island effect and reduce overall transportation emissions. While it is not the only climate intervention in the city, the intended size of the overall project (503 superblocks covering the entire city) and its combination of adaptive measures and mitigation efforts makes it indicative of transformational climate efforts from the side of the municipal authority. It is a case of municipal land use policy with transformational intent that has been integrated into climate policy in Barcelona – this status means that it provides important elements and lessons for transformational climate adaptation.

Our analysis is based on fieldwork conducted in the initial stages of the Poblenou superblock deployment, between March and August 2017, including 13 in-depth, semi-structured interviews with key stakeholders involved in the Barcelona Mobility Plan, in the design and implementation of the superblock, and the mobilisation opposing the project. Interviews were complemented with 17 formal direct observation sessions at the superblock building site, which served to contextualise perceptions and understand social dynamics taking place in and around the superblock. We also complemented fieldwork with an analysis of relevant public plans, local policy documents, media reports (up to July 2019) produced since the initial discussion and visioning work for the superblock, and recent reviews of media articles and informal discussions with local stakeholders working on the further implementation of the superblock intervention through the city. Data was analysed using thematic analysis clustering around phases of the conflict; characteristics and actors of the conflict; resistance discourses; drivers and motivations behind resistance; and resistance responses from municipal stakeholders. In the following section, we elaborate more on those transformational and climate change adaptation characteristics of superblocks.

Throughout this chapter, we present Barcelona's superblocks as a project with transformational qualities that became linked with climate adaptation efforts and is going through a negotiated process of implementation relative to the everyday politics of land use in the city. We argue that this case allows for a clear view of why urban transformational adaptation can only be fully specified from a perspective that accounts for process and outcome at once. In this case, conflict over urban climate adaptation planning of a transformational nature revolves around debates over competitive urbanism and related short-term political gains and this conflict shapes to a great extent environmental benefits that are ostensibly the target of interventions. This implies that, when applied to broader scales, transformational adaptation is likely to be a struggle involving political ideology, urban development, market forces and globalisation, as well as political transitions (such as self-determination in our case) because of the nature and scale of the changes required. We also highlight how civic and political contestation over the authority of *climate champions* can jeopardise not only the potential of transformational adaptation outcomes, but also the political survival of champions themselves. As a result, what is at stake in those struggles goes much beyond environmental planning and policy challenges and disagreements.

The superblock in Poblenou is part of the city of Barcelona's broader ambition to radically transform its mobility, public space, and environmental impact by constructing over 500 superblocks that cover all areas within its boundaries. Physically, a superblock is a traffic-regulated cell of city blocks approximately 400mx400m, which consists of nine smaller blocks in a three-block by three-block mesh.[4] In the outer streets, buses and car traffic circulate, while the newly created space in the interior is reserved mainly for pedestrians and cyclists. Thus, the model allows traffic for residents, services, or emergency vehicles, and for future public space interventions. The plan of creating superblocks in Barcelona first emerged in the 1970s. Early implementations involved traffic pacification plans, which were narrowly designed urban interventions that were not actually called superblocks. The first block free of motorised traffic was set in place in 1993 in Barcelona's old town (Ciutat Vella).[5] Other implementations followed in the 1990s and 2000s, the most well-known in the Gràcia district in 2005. Importantly, the Gràcia traffic pacification programme was not embedded in a larger city-wide mobility plan as was the case with Poblenou, making Gràcia's implementation less articulated into a new vision for the city's development and Poblenou more of a transformational land use and adaptation pilot intervention.

While the boldness of the project has attracted enthusiastic media attention locally (Ara.cat 2018), in Europe (The Guardian 2016), and North America (NY Times 2016),[6] the 2016 pilot superblock project in the district of Poblenou (Fig. 4.1) faced resistance from certain political and civil society spheres. Its implementation at the beginning of September 2016 was characterised by vocal criticism of technical and organisational implementation shortcomings and of lack of citizen participation during the planning and design process. A non-legally binding and low turnout (around 25% participation rate) neighbourhood referendum in May 2017 resulted in the majority of participants voting against continuing the pilot project. This result shaped local media coverage and political debate for several months. In what follows, we explain both the climate change adaptation elements of the project, and its transformational intent.

According to recent studies conducted by the government of the Metropolitan Area of Barcelona, Barcelona Regional and the Catalan Meteorological Service, the

[4] For a more specific description, see Barcelona Ecologia, the public agency in charge of the superblock project implementation: http://www.bcnecologia.net/es/modelo-conceptual/supermanzana

[5] Specifically, in La Ribera neighbourhood, Ciutat Vella (Old City) district.

[6] See also: https://www.youtube.com/watch?v=ZORzsubQA_M

Fig. 4.1 The Poblenou superblock. (Source: Kai Klause)

city of Barcelona is particularly vulnerable to climate change related threats, specifically sea level rise and floods, increasing temperature including urban heat waves, the loss of biodiversity, and more frequent and intense drought periods.[7]

To counteract climate threats, the municipal government follows a cross-scale interconnected action strategy for the Barcelona Metropolitan Area under the umbrella of the Climate Plan (2018–2030). A key early aspect of the city's mitigation and adaptation efforts has been Barcelona's large-scale 2013–2018 Urban

[7] http://lameva.barcelona.cat/barcelona-pel-clima/ca/pla-clima/resiliencia-i-adaptacio

Mobility Plan (PMU 2014), in which the superblock programme is institutionally anchored in different municipal sectorial plans and commitments (Barcelona City Council 2012, 2013, 2014, 2015), and is strategically embedded in a larger vision for the city. In its ambition to drastically reorganise urban mobility infrastructure and land use, the superblock intervention aims at reshaping the city into 503 superblocks and hence reduce car traffic by 21% while restructuring the public transit and cycling system and infrastructure. Environmental goals include diminishing alarmingly high noise levels (in Barcelona, these levels are similar to megacities such as Beijing or Mexico City (World Economic Forum 2017)); reducing the 3500 premature deaths per year[8] associated with air pollution (Pérez et al. 2009; Vert et al. 2017); and, converting a substantial portion of the 60% of space occupied by car use into not only public pedestrian-oriented but also multiple-use space for leisure (e.g. playgrounds), neighbourhood interaction, and activity (local festivals, etc.).

Reflecting the 2013 Green Infrastructure and Biodiversity Plan for 2020 (Barcelona City Council 2013), Barcelona's Commitment to its climate initiative, and Barcelona's adaptation strategy, increasing green spaces and biodiversity to enhance environmental services (i.e. reduction of the effect of heat islands and water retention) while providing new public spaces is also a major focus of the superblock project. A 2014 City of Barcelona reports highlights:

> In order to reduce the heat island effect [...], the management of public space with the creation of new green spaces, the introduction of plant species that generate large areas of shade and the creation of green corridors in urban Barcelona will be essential (Calvet et al. 2014).

The role of green infrastructure for adaptation efforts is well-established (Gill et al. 2007), particularly when it comes to adaptation to heat waves and the reduction of heat islands (WHO 2016). According to WHO, *urban green spaces are an important driver of cities' long-term resilience to climate change and natural disasters* (WHO 2012). Yet, with only 7 m^2 of green space per inhabitant (in some districts 1.85 m^2), Barcelona lacks green spaces in its urban environment (Barcelona City Council 2013). By establishing biodiversity conservation and the increase of urban green as two primary objectives, superblocks help further advance adaptation efforts, a role which has led to the formal inclusion of superblocks within Barcelona's climate action plans.

In sum, Barcelona's superblock programme is an intervention into multiple dimensions of urban life and structure, and cannot be taken exclusively as policy reaction to climate change. Still, adaptation actions are scarcely ever interventions at only one level, as the multi-dimensional character of climate challenge requires efforts to address risks associated with either multiple stresses faced by certain ecosystems or the multiple effects of climate change (Füssel and Klein 2006). This could explain the integration of superblocks to the city's principal policy initiatives as regards climate action, namely Barcelona's Climate Action Plan and the Barcelona

[8]Those figures referred to earlier data; more recent data (2017) show that figures have been reduced down to some 250 persons per year (El Periódico 2017).

Commitment to the Climate initiative. In essence, the superblocks were a vision for transforming the city of Barcelona that both shaped and was shaped by more recent climate adaptation goals.

4.3 Superblocks: A Radical Intervention that Provides a Window into the Everyday Politics of a Transformational Adaptation

When considering the definitions of transformational adaptation provided in our introductory sections, several elements of the superblocks initiative establish its transformational nature. First, the initiative is an intense, non-linear intervention at an enlarged (citywide) scale that aims at reordering the city's mobility systems, and drastically reordering Barcelona's urbanism. Given that 503 superblocks covering the entire city are planned for implementation, this is a large-scale, major project that would transform Barcelona intensely in physical, social, economic, and environmental terms by, as stated by the city government, *radically changing the city (IV5, Executive, Barcelona Urban Development Agency 2017)*. The scale and structure of the project conform with a transformational adaptation initiative.

Second, the superblock project combines climate mitigation and climate adaptation goals. With the superblock project at its core, Barcelona's Urban Mobility Plan 2013–2018 (PMU) is a key element of Barcelona's climate action strategy due to the expected impact of turning car-occupied streets into public green spaces at a large scale, creating bike lanes within and around the superblocks (thus promoting alternative, clean mobility through new infrastructure), and improving local bus networks. Given that the strategic framework of the superblock is also based on Barcelona's Commitment to Climate and its Green and Biodiversity plans, the project is a striking example of a hybrid climate intervention which combines mitigation and adaptation efforts: It seeks to mitigate emissions by adapting the city model to climate change induced threats, while seeking to reduce vulnerability and increase resilience. This view is shared by the Executive of Barcelona's Urban Development Agency, which developed the superblock model and planned its implementation (IV5, Executive, Barcelona Urban Development Agency 2017). Consequently, Barcelona's Climate Plan recognises the extension of the superblock until 2030 as an important measurement for its strategic axes 'Mitigation', 'Adaptation and Resilience', and 'Climate Justice' and stresses both, the superblock's mitigation and adaptation efforts. The project aims at cutting down CO_2 emissions per capita by 40% and reducing climate change-induced heat island effects, or localised temperature spikes (Martín-Vide et al. 2015).

The intention to use superblock green spaces for reducing the heat island effect signifies an adaptation initiative, whereas the intention to reduce the city's contribution to CO_2 emissions points to an initiative aimed at mitigating climate change. This dual environmental purpose is a key characteristic of transformational adaptation (Ribot 2011, IPCC 2014 quoted in Pelling et al. 2014: 2). The superblock is also a type of adaptation recently described as *valuable* in the 2014 IPCC report (p. 25), namely a class of *adaptation options that also mitigate emissions [which] can provide synergies and cost savings in most sectors and system transitions, such as when land management [see: superblock] reduces emissions [see: traffic reduction with superblocks] and disaster risk [in the case of Barcelona superblocks: reducing the risk of high mortality during heatwave episodes by reducing the heat island effect with urban greening provided by the superblock]*. In short, because superblocks blur the line between adaptation and mitigation and integrate profound land use and policy changes at a large scale, the initiative moves closer to transformational status, especially since mitigation seeks to address the generating forces of climate change (e.g. carbon emissions).

Third, while incremental urban adaptation might solely aim at adapting to the consequences of climate change, transformational adaptation actions like the superblock go further and seek to produce fundamental alterations within the urban system that itself produces climate change vulnerability. The municipal government stresses that it seeks a *new urban model* (Barcelona City Council 2017), and the superblock project questions the current paradigm of urban development based on car hegemony over public space. According to the Urban Ecology Agency of Barcelona, black asphalt and car emissions bear the majority of responsibility for urban heat islands (Agencia de Ecología Urbana de Barcelona 2016) and transportation is a major source of greenhouse gas emissions in cities. By cutting down CO_2 emissions drastically, reordering the urban mobility system, and turning streets systematically into green, public space at a large scale, the project aims at radically changing the underlying logic that produces increased rates of climate change vulnerability both in the city and elsewhere.

Overall, the superblock intervention combines the core characteristics of transformational adaptation as defined above based on existing literature: climate change serves as the umbrella goal for organising a series of smaller agendas focused on pollution and health, traffic pacification, noise reduction, and quality of life to produce a non-incremental alteration to the social and mobility system of the city and to profoundly alter land configurations and use. As well, the initiative is framed as a measure that reduces socio-environmental impacts on the most vulnerable social strata by combining adaptation to and mitigation of climate change (BCN Ecologia 2017), and by *radically changing the city* (Executive, Barcelona Urban Development Agency) and its land use. Admittedly, the superblock initiative is not explicitly called a 'transformational adaptation' project, but this in part reflects the fact that the term itself is mostly used in academic circles and is not one yet extensively used by policymakers. Still, this is a multi-dimensional project that serves as a striking example of an action originally *intended to address other problems that can become transformative climate change adaptation* (Kates et al. 2012: 7156). Policymakers do explicitly refer to superblocks as 'transformational' and 'radical', and superblocks have been integrated into the city's climate action narrative.

Given its positioning in Barcelona's climate and land use agendas, this is a strong case for viewing the everyday politics that shape the early implementation of interventions with explicit transformational urban adaptation goals. As the superblock intervention questioned the car-oriented model of urban development in Barcelona by ceding roads to public open space, a complex set of political forces mobilised both for and against the changes. This early period of making transformation happen has highlighted the necessity of developing a process-outcome view of the characteristics of transformational initiatives. Throughout, those with the intention to block the project did so by challenging the authority of the current administration, and thus reshaping the nature of the initiative. The extent to which superblocks will, in the end, be a transformational adaptation in Barcelona depends very much on the interaction between the initiative and these ongoing everyday politics. In what follows, we examine the main aspects of the barriers that arose in a pilot implementation. As a novel contribution to the literature on urban adaptation planning and on transformational adaptation, we focus on the ways in which localised struggles for authority serve as a synecdoche of larger institutional barriers to transformational adaptation. We argue that this particular dimension of process – how struggles for authority are managed – is an additional characteristic of transformational urban adaptation not highlighted in the literature.

4.4 Everyday Struggles for the Seat of Authority

The debate over the Poblenou superblock reveals four aspects of local politics that shape battles over authority, which in turn influence the success of implementing transformational urban policies. First, it demonstrates the fragility of formal municipal authority, especially in the context of efforts to create radical change within a pluralist governance system. Second, it shows that the struggle over who gets political credit is a crucial factor that can determine whether a transformational project will go ahead or not. Third, it illustrates the importance of understanding resistance to 'radical' urban transformation agendas within their broader economic and political contexts, namely the context of a city in the epicentre of projects of economic growth and political independence. Finally, it highlights the importance of linking back controversies over implementation policy (public participation, in our case) to the context of political battles for authority. Overall, it is much more than a fight over different visions for environmental policy and planning. Our analytical points are summarised in Table 4.1 below.

4.4.1 The Fragility of Municipal Authority

As already explained, early 1990s and 2000s superblock interventions were originally conceived in the context of traffic pacification plans and involved individual interventions that were not embedded in a larger city-wide mobility plan – as was the case with the Poblenou superblock.

Table 4.1 Political challenges for transformational adaptation in Poblenou

Political challenges for transformational adaptation	Illustrations from Poblenou Superblock pilot
Overcoming fragility of municipal authority	Countering resistance against the new mayor (Ada Colau), and her non-traditional, grassroots-based political party Fighting opposition to new policies that question real estate speculation, business-centred decisions, and a growth-oriented model of urban development
Gaining political credit	Constructing a vision for Superblocks that moves away from an initial framing of it as a project supporting (automobile) businesses Differentiating oneself from previous centre-right governments and imprinting new types of left-wing policies
Controlling economic agenda and political status of the city	Defining and branding the new Barcelona city model Developing municipal policy centred on residents' quality of life and access to an affordable and liveable city Making decisions for Barcelona as a *city* with specific environmental needs within a broader context of conflict over Catalonia's independence
Supporting civic participation and envisioning the future of the city	Prevailing over perceptions of lack of transparent, prior communication among the local business community Overcoming perceptions of a project controlled by technocrats (e.g. municipal technicians, and outside experts) and *desk intellectuals*, and of a move away from long-term, citizen-driven urbanism Surmounting ideological differences over the city's future growth and development

Over the years, the motives and political orientation of traffic pacification policy changed, reflecting a transition in Barcelona in which 20 years of municipal leadership by socialists (1979–2011) was followed by the rise to power of centre-right parties (2011–2015). Despite being on the desk of municipal governments for a few decades, the superblock project was not approved until the 2015 Urban Mobility Plan (PMU 2014) was passed under Mayor Xavier Trias (2011–2015). Trias' proposal for mobility and public space was mainly oriented toward encouraging economic growth. In line with the 'smart city' model (Eurescom 2018), it aimed at using technological innovation to attract high end developers and residents through provision for electric vehicles – and their growing market – and for continued routes for private cars alongside public transit and bikes as part of a diverse strategy for increasing urban mobility choices (PMU 2014; La Vanguardia 2014a). Trias was a member of the former centre-right party *Convergència i Unió* (CiU), which traditionally stood for business-friendly policies and gained the trust of local economic sectors to secure the city council's approval for the plan.

In contrast, the Poblenou superblock was implemented by the left-wing mayor Ada Colau who was elected in May 2015 (See Fig. 4.2). Known for her activist past as a founding member and public face of the campaign against residential evictions

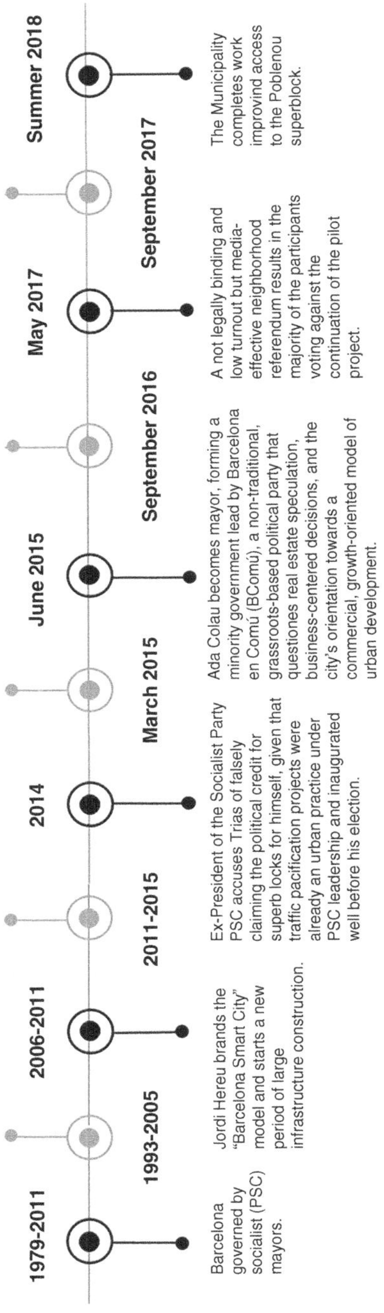

Fig. 4.2 Timeline for Poblenou superblock project

in Spain,[9] Colau was elected on the ticket of a citizen initiative set up to contest municipal power from traditional political parties. Colau's election marked the first time in the city's history since the transition to democracy in 1978 that municipal elections were won by a non-traditional, grassroots-based political party, headed by a female leader who questioned real estate speculation, business-centred decisions, and the city's orientation towards a commercial, growth-oriented model of urban development.

But while Colau won the 2015 municipal election backed by popular enthusiasm, her party *Barcelona en Comú* (BComú)[10] only managed to control 11 out of a total of 41 seats in the city council. This meant that any issue likely to divide public opinion in the city – such as the Poblenou superblock proved to be in 2016 – held potential for opposing parties to make political gains against the mayor and her party.

The vulnerability and threats faced by the mayor and her new party were reflected in the way the Colau Administration approached the superblock project. From the start, there was pressure from the city's administrative directors for agency staff to successfully – but also quickly – implement the project, as recalled by an elected official from the municipality:

> We do have the pressure [to implement the project within the mayoral term]. [...] And yes you have political guidelines, because in the end it is a political question, and this is renewed every four years (IV10, Executive Officer, Department of Urban Model Prospective, Municipality of Barcelona 2017).

Those pressures were then channelled into a struggle for political credit, which influenced reactions to superblocks all through the implementation of the project.

4.4.2 The Struggle for Political Credit

Meanwhile, as the superblock project came to represent a high-profile, radical transformation on a municipal scale linked a new city model, several municipal political groups sought to claim it as their own, and make it serve their purposes. This push began several years before implementation when former mayor Xavier Trias inserted superblocks into the city's Mobility Plan, framing the intervention at first as one that would serve business interests. In response to Trias' initiative, the leader of the opposition Catalan Socialist Party municipal group accused Trias of falsely claiming the political credit for superblocks for himself (La Vanguardia 2014b), given that traffic pacification projects were already an urban practice inaugurated well before his election.

[9] Also known as Platform for People Affected by Mortgages (PAH from its initials in Spanish, standing for *Plataforma de Afectados por la Hipoteca*), see: http://afectadosporlahipoteca.com/

[10] Barcelona in Common – in Catalan *Barcelona en Comú* (previously known as *Guanyem Barcelona*).

Similarly, Trias' political group accused the BComú municipal government of trying to claim credit for political innovation through the implementation of the Poblenou superblock. Just as Trias was said to have co-opted the idea, his party said Colau co-opted the idea. Thus, to increase its ownership of the superblock initiative, instead of launching the Poblenou pilot project as it had been planned over a long period by the previous municipal government, BComú opted for creating a new proposal from scratch just a few blocks away. For Poblenou superblock opponents, that decision meant that the ability to claim project ownership and differentiate oneself from previous centre-right governments was prioritised over careful planning:

> Colau could have implemented [this planned superblock] [...] if she had wanted to. And why didn't they do this? Because everything had already been done. So, this way they couldn't have said that *it was them* [stress added] who did the superblocks! Because [those] who developed it were the ones who were [in power] before. But it's a political issue of 'no, no, this [project] was mine (IV4, Former CiU councillor 2017).

In the minds of many of those involved at the time, there was a lot at stake with regard to where the credit for the superblock project went. Project success might ensure an indelible new political heritage in Colau's favour. If a transformation of this nature, which some linked as far back as the unrealised vision of Barcelona's twenty-first century acclaimed planner, Ildefons Cerdà, was accomplished by Colau, then it would certainly burnish her image and very possibly mean that her citizen-centred rather than Trias' business-centred vision for a green city would prevail in the long term. Given the scope and media attention afforded to the project, such a victory would be accompanied with international legitimacy for the current Administration. In that sense, the progressive agenda of BComú represented a risk to the political establishment:

> long-established [forces], which cannot tolerate that their power is taken away. So, they have to try to annihilate it [the threat to that power] (IV5, Executive, Barcelona Urban Development Agency 2017).

As a result, the local political opposition saw high stakes in contesting who got credit for what. Before the Poblenou superblock had a contentious public identity, both those for and against the Colau Administration jockeyed for the right to claim ownership over the transformational aspects of the project and (re)gain the lead position in superblock implementation. This struggle for appropriation led to even more intensive political polarisation, where:

> all right-wingers [were] up against the superblocks because they have been the proposal of BComú (IV13, Local Businessperson 2017).

Such polarisation coincided with disagreements on a number of key matters in both the economic and political realms, as we explain in the following section.

4.4.3 A Contestation Over the Economic Agenda and Political Status of the City

Those tensions were coupled with a fear among more conservative interests that the project represented a wider attempt to push an anti-economic growth agenda. Some residents termed the BComú approach as *fanatical radicalism*, (El Nacional 2017) which they saw as rejecting economic growth. The paramount concern of those opposition parties was to prevent BComú from defining and branding the new Barcelona city model in ways that might threaten certain historic business and economic interests and the privileged access of those interest groups to local government through established social networks. This concern, indeed, extended far beyond the design and implementation of superblocks to other early decisions from Ada Colau, such as cancelling municipal subsidies for the Formula 1 Barcelona circuit (Marca 2015) and banning licenses for the construction of new hotels (El País 2015). Superblocks became for many a means of addressing these wider concerns. The ex-spokesman of the previous municipal government party (CiU), Joaquim Forn, expressed this point in 2016 (El Periódico 2016):

> The superblock became a *botched job* thanks to *[t]he anti-private vehicle obsession of Ada Colau's government.*

Claims such as these indicate how criticisms were as much against an agenda which included developing municipal policy centred on residents' quality of life, access to an affordable city, anti-mass tourism, citizen-driven public spaces, and non-traditional models of local economic development,[11] as they were against the superblock. Ada Colau has indeed been vocal in her rejection of a Barcelona city model tailored to the interests of entrepreneurs, star architects, real estate investors, and tourist industry business owners, a city model much criticised in the past (Capel 2005, 2007; Delgado 2007; Montaner 2004). However, the characterisation of her actions as an *anti-private vehicle obsession* was representative of the hyperbole used by her opponents to describe these positions as dangerously radical and untethered to reality. As a result, the superblock conversation shifted from a battle over credit to one where the initiative was portrayed as an extreme expression of the dangerous radicality of the Colau Administration. Much of the opposition, which had previously supported and voted for the project, saw it as an existential threat.

Moreover, left-wing parties other than BComú, such as the independentist Republican Left of Catalonia (ERC)[12] did not rally to support the project (Catalunyapress 2017). Rather, the political vulnerability around superblocks that right-wing parties had manufactured became also relevant for left wing debates over independence for Catalonia. Here one needs to keep in mind that Barcelona is the

---

[11] See Colau's interview with Al Jazeera in 2017: https://www.aljazeera.com/programmes/talktojazeera/2017/04/barcelona-mayor-city-losing-identity-170412082645192.html and Barcelona en Comú's 2015 program: https://barcelonaencomu.cat/sites/default/files/programaencomun_cast.pdf

[12] From its initials in Catalan: *Esquerra Republicana de Catalunya* (Republican Left of Catalonia).

largest Catalan city and the capital of Catalonia, which goes some way in explaining its economic importance and political status. Left wing parties that favoured independence of Catalonia from Spain did not see in Ada Colau a partner for their cause – from which Colau separated herself although she politically and materially supported a referendum for independence. Since May 2016, when Colau moved into a pact with the Socialists for governing the City Hall, ERC became less cooperative to the projects of a mayor who closely collaborated with a party (Socialists) that would not support Catalan independence.

In those ways, superblocks became a symbol for deeper ideological divides over who should benefit from Barcelona's growth model and got indirectly embroiled in the question of whether Catalonia should be independent – much more than disagreements over sustainability, climate, or environmental policies at large. As a result, societal opposition against the project partly emerged or became heightened because those who did not agree with the ideological orientation of the party in the municipal seat of power were against any project they perceived as being connected to it and the mayor.

> The government of Colau is a government of passion for and against it. […] And now it's not 'the superblocks of Barcelona', it's 'the superblocks of Colau'. [...] Those who are against the party, government or the person itself are against the project (IV6, City councillor, ERC 2017).

Therefore, the more transformational the project, the more opposition it gathered, and transformation translated into political vulnerability. The right way to address climate mitigation and adaptation and the best mode for increasing urban mobility had very little to do with this opposition.

4.5 Participation Complaints and Clashing Visions of the City

Last, the fight over the seat of authority was linked to a public conversation over participatory democracy associated with the Poblenou Superblock. Perceptions of exclusion from the implementation process were the most often-reported reasons for neighbourhood residents and businesses to reject the project. Importantly, exclusion complaints were tied into the fundamental, ideological differences over the city's development vision. In their own words, many residents and local business owners were taken by surprise as the superblock was inaugurated in the middle of the 2016 summer vacation without any previous warning or announcement:

> We deny that a participatory process has happened. [...] [At the beginning of September] some placards began to appear in our neighborhood, but the municipality did not explain the superblock. They explained it eight days later more or less [after implementation had happened]. […] If there were a real open participatory process, surely […] the acceptance would have been much bigger. [That] would change the situation for businesses as well (IV8, Local Neighborhood Association Representative 2017).

Owners of car shops and automobile repair shops were particularly taken-aback by the implementation of the superblock and became vocal early on in Fall 2016 about their opposition to the project. They argued that the project was not implemented with transparent prior communication about its possible business impacts and showed insufficient protection for established economic activities in the neighbourhood – especially car dealers and mechanics:

> We are an automobile repair shop, we make fast repairs of vehicles. [Before], many people passed [on their way]... to the centre of Barcelona. [...] This traffic has been greatly reduced. Also billing levels are falling (IV13, Local Business Owner 2017).

But beyond implementation shortcomings, exclusion complaints also expose fundamental, ideological differences over the city's growth and development vision, linking back to clashing visions of and for the city. In that context, local automobile-oriented businesses became a centrepiece of the opposition adding strength to wider efforts from pro-growth interests and the pro-independence left to undermine the Administration with critiques of the participation process. Car dealers and mechanics, and a cohort of local residents strongly endorsed the view that the superblock's final design and location was imposed by technicians, outside experts, and *desk intellectuals* advancing a radical intervention rather than the incremental and flexible project that local residents preferred. Embedded in that criticism of top-down implementation, the neighbourhood referendum was meant to corrode the ideological stance of BComú and challenge its commitment to citizen-driven urbanism:

> The referendum is a political lever [...]. [It] also serves to put in contradiction the fundamental speech of Colau. Grassroots participation, referendum for everything, it seemed like Switzerland. Now we are expressing a contradiction in her basic discourse (IV8, Local Neighborhood Association Representative 2017).

Local residents and businesses linked exclusion complaints to a more fundamental rejection of transforming mobility patterns in the area. Local opponents argued that the section of Poblenou where the project was deployed is not an area where social interaction and mixed uses in public spaces have ever been valued by citizens, and that local residents had never expressed a need for public space interventions. Simultaneously, in the mind of local opponents, the superblock elicited a cognitive dissonance with accepted socio-cultural habits that are essential to a *modern* city. Participation complaints became in some ways a proxy for the unease raised by this cognitive dissonance. This was a struggle that goes to the heart of what was trying to be transformed:

> For [BComú], the discourse is 'we must recover space occupied by cars' [...]. We don't want them to put all that stuff on the streets, children's games and so on. There is no need to put that in the middle of the street. [...] It's empirically established, since 10,000 years or 20,000 years ago: there is more economic activity in places where people pass through (IV8, Local Neighborhood Association Representative 2017).

Here, one cannot overlook that the image of Barcelona as a city with remnants of rural and so-called *underdeveloped* characteristics in the 1970s and 1980s is still present in the imaginary of many residents and political leaders, even outside the context of the superblock. Thirty to forty years ago, Barcelona still had informal settlements on the waterfront and outskirts of the city, called *barracas*[13] (Andrés Creus 2011) and the presence of donkeys and chicken on the streets was a sign of a not-so-remote *backward* (Álvarez et al. 2017) past which many want to forget. The stance of many superblock opponents reflected a rejection on how authorities aimed at defining how to build a *modern*, developed, and liveable city. And of course, opponent imaginaries of the city tied quite well with those defended by the mayor's opponents in the struggle for municipal authority:

> We cannot imagine a world in which the private vehicle can be dispensed with (IV1, City Councillor of centre-right liberal political party Ciudadanos 2017).

4.6 Conclusions

This chapter proposes the early implementation of the Barcelona superblocks as a case that highlights the importance of everyday local politics – and specifically struggles for authority – in determining whether an initiative with transformational adaptation goals ever achieves that status. It responds to knowledge gaps around the role of political barriers and power relations in transformational urban practice by detailing how and why transformational adaptation is not only about the character of the initiatives themselves but also the period of making transformation happen. In all, we have inquired as to which barriers emerge and how, when authority tries to pursue transformational action at the city level and faces resistance, and what these political dynamics tell us about best ways of implementing transformational adaptation at the urban level.

Having in mind that urban adaptation practices are still lagging behind in complying with the increasing academic calls for transformational approaches, this case study shows that the underlying drivers of barriers to urban transformational adaptation initiatives are to be found in political struggles for authority. As our analysis demonstrates, challenges to implement urban climate adaptation planning of a transformational nature are more driven by competitive urbanism and related short-term political gains than by the contested importance of environmental benefits framed by those interventions. Because of the scope of changes needed, transformational adaptation is a fight involving political ideology, urban development, market forces and globalisation, and political transitions and self-determination more than it is a fight over different visions for environmental policy and planning.

In short, we argue that *resolving local struggles for authority* is a necessary condition for transformational urban adaptation that is not explicitly labelled as such in the literature. In an outcome-only view of urban transformational adaptation, this process-oriented quality is left off. However, the case analysed here shows that it is

[13] http://www.lavanguardia.com/libros/20111109/54237962987/barracas-de-barcelona-memorias-contrastadas.html

not less central to the transformational capacity of an urban intervention than the other more outcome-oriented qualities that have been a focus of the literature. This is the main lesson of the Barcelona case. In order for an initiative to move beyond incremental approaches, the initiative itself must have transformational qualities, but must also be accompanied by a transformational process of resolving local struggles for authority.

Our analysis of the pilot superblock in Poblenou shows that the contestation of authority that materialises in clashes over profound ideological differences and deeply held norms are key elements of the multi-scalar political nature and challenges facing transformational urban adaptation. This in turn tells us something about the factors specific to urban adaptation to climate change that shape urban politics and power struggles. In particular, a major challenge of implementing transformational adaptation strategies relates to the non-linear, radical and transformational character of those interventions. This element turns transformational adaptation to a type of urban planning that challenges established development pathways by questioning the very essence of a system, including the political economic structures that underpin urban development. Hence, it inevitably comes up against issues of power and is rarely met with political consensus. Rather, implementing a project like the superblocks creates an ideological tug-of-war in which the power to frame and enforce interpretations of the city model and visions of the future city is fought out. In that sense, urban transformational adaptation must confront unavoidable socio-political and structural barriers and is at least as much about the fight for power as about transforming material relations, or environmental issues themselves. Far from being solely about the best way to address climate change, environmental policy, or urban resilience planning, political barriers to transformational adaptation involve a structural battle over who has the right to design and brand the new model for the city – and who can and/or will harness political gains from it.

In agreement with Kates et al. (2012) we find that behavioural actions that tend to maintain existing resource systems and policies made it difficult to implement the superblock transformational initiative. We also find that power, politics and interests have posed considerable barriers (Pelling and Manuel-Navarrete 2011), and that the superblock case showcases that even *well-intentioned governments may be too structurally dependent on the interests of capital and too electorally dependent on the support of the groups who either already have, or aspire to, high consumption and fossil fuel-dependent lifestyles* (Godfrey-Wood and Otto Naess 2016). But beyond those rather general observations about the links between power and transformational adaptation, the implementation of superblocks tells us that, as with climate change itself, climate change adaptation does not take place in a vacuum with relation to everyday life and other political priorities unrelated to *the environment*. It happens, as the Barcelona case reveals, while people continue with their *business-as-usual* activities, which include running a business and commuting by car; engaging in political manoeuvring (including efforts to acquire political credit) in the effort to get hold of or maintain institutional power; and, pursuing, or frustrating other political projects (e.g. independence in the Catalan case).

Transformational action to adapt to climate change should pay heed to such everyday obstacles, which relate not only to ways in which authority is either imposed upon or internalised by subjects (e.g. evident in reactions to change deep-seated habits such as reducing car use, and in the tendency to construct images of neighbourhood as a space primarily for commercial activity), but also with the practical politics of the hunt for the seat of authority, for acquiring the supreme, legitimate right to decide for *the collective* (the city and its citizens in this case) and the vision for its future. Here, our findings reflect previous urban climate adaptation research underlining the role of climate champions for pushing forward adaptation agendas and interventions, such as those in Quito and Durban, where climate leaders managed to integrate adaptation interventions into sustainability practices and build on existing successes and capacities (Carmin et al. 2013). They are also in line with recent research in Durban and Indore which argues that adaptation actions must be led considering the powerful, and often entrenched, political economic forces and interest that shape urban development (Chu et al. 2017). However, our chapter is novel in highlighting how civic and political contestation of champions' authority to decide can jeopardise transformational adaptation achievements – and the political survival of champions themselves.

Indeed, the entanglement of climate change adaptation in struggles for authority in Spain is not an isolated characteristic of the Barcelona superblock case. In August 2017, the Catalan regional parliament passed the Catalan Climate Change Bill into law (The Climate Group 2017). Although hailed internationally for its ambition and boldness (The Climate Group 2017), the response of the Spanish government was to challenge the law in court. The rationale for doing so was essentially (and inter alia) that conceding to the Catalan parliament the capacity to legislate for issues related to climate change would imply accepting the taking over of state competences and acknowledging to Catalonia legislative capacities that correspond to an independent state (El País 2017). Although the Spanish Constitutional Court ruled against the Spanish government (El Periódico 2017), the case serves to highlight how attempts to respond drastically to climate change challenges (the Catalan law aspires to make Catalonia 100% carbon free by 2050) in Spain are entangled with struggles around capturing or holding the seat of legitimate authority, even more so now with the political battle over the independence of Catalonia. Once again, this suggests that large-scale and deep transformations can be obstructed not only out of fear for the actual material and political effects of transformation per se, but also because of the message they convey as concerns who has the authority to decide for *the common good*.

As concerns implementation lessons at the urban level, we find that the key barriers that emerged in the Barcelona superblock project were related not only to the political struggle for capturing municipal authority, but also to discontent over public participation aspects of project implementation and a perceived imposition of authority. Planning initiatives that aspire to alter urban spaces along transformational adaptation lines should try to foresee how to engage with such reactions during project implementation. In that sense, our findings coincide with other calls for genuine participation processes (Gillard et al. 2016) and for creation of spaces of engagement and negotiation (O'Brien et al. 2015). Nevertheless, our findings provide mixed insights as concerns the alleged capacity of pluralism to produce *more ambitious and politicised visions of the future*, and the potential of discord and instability in micropolitics to provide *potent opportunities for innovation and change to emerge* (Gillard et al. 2016), as in the Barcelona superblock case, the presence of pluralism ended up blocking a transformational initiative. This should not be taken as advocacy for authoritarian green politics or collective decision-making in general, but as a reminder of the contradictions of pluralism in search of sustainable and resilient solutions. Relatedly, the prioritisation by local shop owners of potential economic losses from the implementation of superblocks is in line with Pelling et al. (2015) who explain that in participatory methodologies there is a tendency for communities to prioritise immediate risks or needs.

The superblock case puts in evidence that urban transformational adaptation must confront unavoidable socio-political barriers and allow for the construction of coalitions that can build together common imaginaries about the future of the city. The evolution of the initial superblock conflict during 2017 and 2018 and the greater public acceptance of the superblocks[14] also shows that, as transformational land-use planning later integrates more defined benefits or features (i.e. playgrounds, trees and plants, street infrastructure) and as those provide new social benefits, the mid-to long-term evolution seems to strengthen the ability of transformational interventions to be operationalised on the ground. This is an important lesson for other cities planning to conduct such structural land use changes for transformational adaptation – brave politics in the short-term is needed to achieve mid- to long-term transformational goals.

In that sense, other cities should also take into consideration that urban transformational adaptation is more about the fight for urban political power – as related to city vision, stakeholder participation and governance, and the distribution of the material benefits of economic modernity – and competitive urbanism at the local and global scales, than about environmental issues themselves. As of summer 2022, the municipality has deployed several new superblocks and green axes, each of them with its own urban development and design challenges – the main second superblock was completed in early 2020 in the neighbourhood of Sant Antoni amidst concerns over gentrification and tourism massification in downtown Barcelona. The aftermath of the 2019 municipal elections – through which Ada Colau is governing in a coalition with the PSC socialist party *but* also with the other left (independentist) party ERC in the opposition – is placing such interventions at the centre of new political struggles for (future) authority and for deciding the urban future of the city. Although it appears that new Superblocks and their associated green axes are facing lesser public resistance and greater political and societal acceptance, civic groups such as *EnsPlantem* and the *Observatori dels Barris del Poblenou* are concerned about land and real estate speculation and housing price increases in the vicinity of the Superblock and its adjacent green projects. Therefore,

[14] https://cat.elpais.com/cat/2018/09/29/catalunya/1538246791_684437.html

for transformational adaptation to be environmentally and socially sound as well as more broadly accepted, it must place social justice and housing rights at the centre of urban planning.

Acknowledgements The authors would like to thank Miquel Ortega for comments on an earlier draft of this analysis. We also acknowledge financial support from the Ramón y Cajal Programme (Contract number: RYC-2015-17372), the Juan de la Cierva Programme (IJCI-2016-31100) and the Maria de Maetzu Centre of Excellence Programme (MDM-2025-0552) funded by the Spanish Ministry of Science, Innovation & Universities and the European Social Fund; as well as the European Research Council Starting Grant GreenLULUs (GA678034).

References

Álvarez J, Oliva V, Cadena JM (2017) Barcelona desapareguda de Giacomo Alessandro. Barcelona Llibres. Ajuntament de Barcelona, Barcelona

Anguelovski I, Shi L, Chu E, Gallagher D, Goh K, Lamb Z, Reeve K, Shi L, Teicher H (2016) Equity impacts of urban land use planning for climate adaptation: critical perspectives from the global north and south. J Plan Educ Res 36(3):333–348. https://doi.org/10.1177/0739456x16645166

Ara.cat (2018) La superilla del Poblenou, finalista del Premi Europeu de l'Espai Públic Urbà 2018. https://m.ara.cat/cultura/Superilla-Poblenou-Premi-Europeu-Public_0_2012198997.html. Accesssed 05/24/2018

Araos M, Berrang-Ford L, Ford JD, Austin SE, Biesbroek R, Lesnikowski A (2016) Climate change adaptation planning in large cities: a systematic global assessment. Environ Sci Pol 66:375–382

Bahadur AV, Tanner T (2014) Transformational resilience thinking: putting people, power and politics at the heart of urban climate resilience. Urban Environ 26(1):200–214

Barcelona City Council (2012) Compromiso Ciudadano por la Sostenibilidad 2012–2022

Barcelona City Council (2013) Medi Ambient i Serveis Urbans – Hàbitat Urbà. Barcelona Green Infrastructure and Biodiversity Plan 2020

Barcelona City Council (2014) New urban mobility plan of Barcelona 2013–2018

Barcelona City Council (2015) Hàbitat Urbà – Medi Ambient i Serveis Urbans. Plan to improve air quality in Barcelona 2015–2018

Barcelona City Council (2017) Presentació Superilles. http://ajuntament.barcelona.cat/superilles/es/presentacion. Accessed 04/10/2018

BCN Ecologia (2017) Porque supermanzanas. http://bcnecologia.net/es/porque-supermanzanas. Accessed 05/27/2018

Bicknell J, Dodman D, Satterthwaite D (eds) (2009) Adapting cities to climate change: understanding and addressing the development challenges. Earthscan, London

Calvet M, Biosca O, Ulied A (2014) Ajuntament de Barcelona. Anàlisi dels plans d'adaptació al canvi climàtic

Capel H (2005) El Modelo Barcelona: un examen crítico. Eure, Santiago

Capel H (2007) El debate sobre la construcción de la ciudad y el llamado 'Modelo Barcelona'. Scripta Nova – Revista Electrónica de Geografía y Ciencias Sociales 11(233):741–798

Carmin J, Dodman D, Chu E (2013) Urban climate adaptation and leadership: from conceptual understanding to practical action. OECD regional development working papers, 2013/26. OECD Publishing, Paris

Catalunyapress (2017) La oposición se une a los vecinos y exige la retirada de la superilla de Poblenou. http://www.catalunyapress.es/texto-diario/mostrar/745073/oposicion-une-vecinos-exige-retirada-superilla-poblenou. Accessed 05/29/2018

Chu EK (2015) The political economy of urban climate adaptation and development planning in Surat, India. Eviron Plann C Gov Policy 00:1–18. https://doi.org/10.1177/0263774X15614174
Chu EK (2017) Urban climate adaptation and the reshaping of state–society relations: the politics of community knowledge and mobilisation in Indore India. Urban Stud 55(8):0042098016686509
Chu EK, Anguelovski I, Roberts D (2017) Climate adaptation as strategic urbanism: assessing opportunities and uncertainties for equity and inclusive development in cities. Cities 60:378–387
Connolly JT (2018) From systems thinking to systemic action: social vulnerability and the institutional challenge of urban resilience. City Community 17(1):8–11
de Andrés Creus L (2011) Barraques. La lluita dels invisibles. Pròleg de Pasqual Maragall. Ara llibres, Serie H, Barcelona
Delgado M (2007) La ciudad mentirosa: Fraude y miseria del modelo Barcelona. Libros de la Catarata, Barcelona
EEA (European Environment Agency) (2016) Urban adaptation to climate change in Europe 2016—Transforming cities in a changing climate. EEA, Copenhagen. EEA Report No. 12/2016
El Nacional (2017) Albert Acín. Los críticos de la supermanzana del Poblenou exigen cambios a Colau. https://www.elnacional.cat/es/bcn-hub/supermanzana-poblenou-oposicion-vecinos_131883_102.html. Accessed 04/22/2018
El País (2015) Barcelona mayor introduces one-year ban on new tourist accommodation. https://elpais.com/elpais/2015/07/02/inenglish/1435828213_655223.html. Accessed 05/16/2018
El País (2017) Manuel Planelles. El Gobierno frena el intento de Cataluña de cerrar las nucleares en diez años. https://politica.elpais.com/politica/2017/11/03/actualidad/1509723721_533265.html. Accessed 05/29/2018
El Periódico (2016) Cristina Savall. La 'superilla' divide al Poblenou. https://www.elperiodico.com/es/barcelona/20160912/experimento-superilla-poblenou-provoca-debate-divide-vecinos-5375063. Accessed 05/29/2018
El Periódico (2017) Superar los límites de contaminación causa 250 muertes anuales en Barcelona. https://www.elperiodico.com/es/sociedad/20171129/superar-los-limites-de-contaminacion-causa-250-muertes-anuales-en-barcelona-6458997. Accessed 05/16/2018
Eriksen SH, Brown K (2011) Sustainable adaptation to climate change. Clim Dev 3(1):3–6
Eriksen SH, Nightingale AJ, Eakin H (2015) Reframing adaptation: the political nature of climate change adaptation. Glob Environ Chang 35:523–533
Eurescom (2018) Becoming smart is a transitional process. Interview on Smart Cities with Xavier Trias, former mayor of Barcelona. https://m.eurescom.eu/news-and-events/eurescommessage/eurescom-message-winter-2015/becoming-smart-is-a-transitional-process-interview-on-smart-cities-with-xavier-trias.html. Accessed 05/10/2018
Fairhead J, Leach M, Scoones I (2012) Green grabbing: a new appropriation of nature? J Peasant Stud 39:237–261
Fieldman G (2011) Neoliberalism, the production of vulnerability and the hobbled state: systemic barriers to climate adaptation. Clim Dev 3(2):159–174
Füssel HM, Klein RJ (2006) Climate change vulnerability assessments: an evolution of conceptual thinking. Clim Chang 75(3):301–329
Gill SE, Handley JF, Ennos AR, Pauleit S (2007) Adapting cities for climate change: the role of the green infrastructure. Built Environ. Alexandrine Press 33(1):115–133(19)
Gillard R, Gouldson A, Paavola J, Van Alstine J (2016) Transformational responses to climate change: beyond a systems perspective of social change in mitigation and adaptation. Wiley Interdiscip Rev Clim Chang 7(2):251–265
Glavovic BC (2008) Sustainable coastal communities in the age of coastal storms: reconceptualising coastal planning and ICM as 'new' naval architecture. J Coast Conserv 12(3):125–134
Godfrey-Wood R, Otto Naess L (2016) Adapting to climate change: transforming development? IDS Bull 47(2)
Hadarits M, Pittman J, Corkal D, Hill H, Bruce K, Howard A (2017) The interplay between incremental, transitional, and transformational adaptation: a case study of Canadian agriculture. Reg Environ Chang 17(5):1515–1525

IPCC (2014) In: Field CB, Barros VR, Dokken DJ, Mach KJ, Mastrandrea MD, Bilir TE, Chatterjee M, Ebi KL, Estrada YO, Genova RC, Girma B, Kissel ES, Levy AN, MacCracken S, Mastrandrea PR, White LL (eds) Climate change 2014: impacts, adaptation, and vulnerability. Part A: global and sectoral aspects. Contribution of Working Group II to the Fifth Assessment Report of the Intergovernmental Panel on Climate Change. Cambridge University Press, Cambridge, UK/New York, p 1132

Kates RW, Travis WR, Wilbanks TJ (2012) Transformational adaptation when incremental adaptations to climate change are insufficient. Proc Natl Acad Sci 109(19):7156–7161. http://doi. https://doi.org/10.1073/pnas.1115521109

La Vanguardia (2014a) Trias apuesta por una reactivación económica con movilidad sostenible. http://www.lavanguardia.com/politica/20140305/54402065089/trias-apuesta-por-una-reactivacion-economica-con-movilidad-sostenible.html. Accessed 05/03/2018

La Vanguardia (2014b) PSC afea a Trias que se apropie de ideas viejas y olvide dignidad de barrios. http://www.lavanguardia.com/politica/20140226/54401773167/psc-afea-a-trias-que-se-apropie-de-ideas-viejas-y-olvide-dignidad-de-barrios.html. Access 05/29/2018

Lang S, Rothenberg J (2017) Neoliberal urbanism, public space, and the greening of the growth machine: New York City's High Line Park. Environ Plan A 49(8):1743–1761

Lonsdale K, Pringle P, Turner B (2015) Transformative adaptation: what it is, why it matters & what is needed. UK Climate Impacts Programme, University of Oxford, Oxford, UK

Lubell M, Feiock RC, La Cruz D, Ramirez EE (2009) Local institutions and the politics of urban growth. Am J Polit Sci 53(3):649–665

Marca (2015) Ada Colau: "Subvencionar la Fórmula 1 no es la prioridad". http://www.marca.com/2015/05/27/motor/formula1/1432744515.html. Accessed 05/29/2018

Marshall NA, Park SE, Adger WN, Brown K, Howden SM (2012) Transformational capacity and the influence of place and identity. Environ Res Lett 7(3):034022

Martín-Vide J, Artola VM, Cordobilla MJ, Moreno M (2015) La isla de calor en el área metropolitana de Barcelona y la adaptación al cambioclimático. Universitat de Barcelona & Barcelona Regional (BR), Metrobs

Montaner JM (2004) La evolución del modelo Barcelona (1979–2002). In: Borja J, Muxí Z, Cenicacelaya J (eds) Urbanismo en el siglo XXI: una visión crítica: Bilbao, Madrid, Valencia, Barcelona, 1st edn, Barcelona, Escola Tècnica Superior d'Arquitectura de Barcelona: Edicions UPC, pp 203–222

Mustelin J, Handmer J (2013) Triggering transformation: managing resilience or invoking real-change? In: Proceedings of transformation in a changing climate conference, 19–21 June 2013. University of Oslo, Norway

Nightingale AJ (2018) The socioenvironmental state: political authority, subjects, and transformative socionatural change in an uncertain world. Environ Plan E Nat Space 1(4):688–711

Nightingale AJ, Ojha HR (2013) Rethinking power and authority: symbolic violence and subjectivity in Nepal's Terai Forests. Dev Change 44:29–51

NY Times (2016) What New York can learn from Barcelona super blocks. https://www.nytimes.com/2016/10/02/nyregion/what-new-york-can-learn-from-barcelonas-superblocks.html?mcubz=0. Accessed 05/29/2018

O'Brien K (2012) Global environmental change II: from adaptation to deliberate transformation. Prog Hum Geogr 36(5):667–676

O'Brien K, Sygna L (2013) Responding to climate change: the three spheres of transformation. Proc Transform Changing Clim 16:19–21

O'Brien K, Eriksen S, Inderberg TH, Sygna L (2015) Climate change and development: adaptation through transformation. In: Climate change adaptation and development: transforming paradigms and practices. Routledge, London

Park SE, Marshall NA, Jakku E, Dowd AM, Howden SM, Mendham E, Fleming A (2012) Informing adaptation responses to climate change through theories of transformation. Glob Environ Chang 22(1):115–126

Pelling M (2010) Adaptation to climate change: from resilience to transformation. Routledge, London

Pelling M, Manuel-Navarrete D (2011) From resilience to transformation: the adaptive cycle in two Mexican urban centers. Ecol Soc 16(2):Article 11

Pelling M, O'Brien K, Matyas D (2014) Adaptation and transformation, climatic change. In: Birkmann J, Mechler R (eds) Special issue on advancing climate change adaptation and risk management. https://doi.org/10.1007/s10584-014-1303-0

Pelling M, O'Brien K, Matyas D (2015) Adaptation and transformation. Clim Chang 133(1):113–127

Pérez L, Sunyer J, Künzli N (2009) Estimating the health and economic benefits associated with reducing air pollution in the Barcelona metropolitan area (Spain). Gac Sanit 23(4):287–294

PMU (2014) Pla de Mobilitat Urbana de Barcelona 2013–2018. Ajuntament de Barcelona. http://mobilitat.ajuntament.barcelona.cat/sites/default/files/PMU2013-2018_DocAprovInicial_Parte1_0.pdf. Accessed 05/27/2018

Revi A, Satterthwaite D, Aragón-Durand F, Corfee-Morlot J, Kiunsi R, Pelling M, Roberts D, Solecki W, Gajjar SP, Sverdlik A (2014) Towards transformative adaptation in cities: the IPCC's Fifth Assessment. Environ Urban 26(1):11–28. https://doi.org/10.1177/0956247814523539

Ribot J (2011) Vulnerability before adaptation: toward transformative climate action. Glob Environ Chang 21(4):1160–1162

Rickards L, Howden SM (2012) Transformational adaptation: agriculture and climate change. Crop Pasture Sci 63(3):240–250

Shi L, Chu E, Anguelovski I, Aylett A, Debats J, Goh K, VanDeveer S (2016) Roadmap towards justice in urban climate adaptation research. Nat Clim Chang 6(2):131–137

Sikor T, Lund C (2009) Access and property: a question of power and authority. Dev Chang 40:1–22

Taylor M (2014) The political ecology of climate change adaptation: livelihoods. agrarian change and the conflicts of development. Explorations in Development Studies, Routledge

Termeer CJ, Dewulf A, Biesbroek GR (2017) Transformational change: governance interventions for climate change adaptation from a continuous change perspective. J Environ Plan Manag 60(4):558–576

The Climate Group (2017) Catalonia passes climate change law to reduce emissions by 100% by 2050. https://www.theclimategroup.org/news/catalonia-passes-climate-change-law-reduce-emissions-100-2050. Accessed 05/29/2018

The Guardian (2016) Superblocks to the rescue: Barcelona's plan to give streets back to residents. https://www.theguardian.com/cities/2016/may/17/superblocks-rescue-barcelona-spain-plan-give-streets-back-residents. Accessed 05/29/2018

Vert C, Sánchez-Benavides G, Martínez D, Gotsens X, Gramunt N, Cirach M, Crous-Bou M (2017) Effect of long-term exposure to air pollution on anxiety and depression in adults: a cross-sectional study. Int J Hyg Environ Health 220(6):1074–1080

Warner BP, Kuzdas CP (2017) The role of political economy in framing and producing transformative adaptation. Curr Opin Environ Sustain 29:69–74

Whitehead M (2013) Neoliberal urban environmentalism and the adaptive city: towards a critical urban theory and climate change. Urban Stud 50(7):1348–1367

World Economic Forum (2017) These are the cities with the worst noise pollution. https://www.weforum.org/agenda/2017/03/these-are-the-cities-with-the-worst-noise-pollution/. Accessed 05/27/2018

World Health Organization (2016) Urban green spaces and health. Copenhagen: WHO. Regional Office for Europe

Ziervogel G, Cowen A, Ziniades J (2016) Moving from adaptive to transformative capacity: building foundations for inclusive, thriving, and regenerative urban settlements. Sustainability 8(9):955

Zografos C (2017) Flows of sediment, flows of insecurity: climate change adaptation and the social contract in the Ebro Delta, Catalonia. Geoforum 80:49–60

Zografos C, Anguelovski I, Grigorova M (2016) When exposure to climate change is not enough: exploring heatwave adaptive capacity of a multi-ethnic, low-income urban community in Australia. Urban Clim 17:248–265

Chapter 5
Urban Resilience in Latin America: Questions, Themes and Debates

Pere Ariza-Montobbio, Andrea Carrión, and Gian Carlo Delgado-Ramos

Abstract Urban resilience is gaining attention as an integral approach to face climate change and other social-ecological stresses faced by cities. These shocking events, unevenly distributed across social groups and regions, aggravate existing structural inequalities that characterise neoliberal urbanisation in the Global South. Accordingly, a critical and interdisciplinary approach has arisen to address environmental justice and the political ecology of urban resilience. Through such a lens, this chapter explores key themes, questions and debates around urban resilience in Latin American and the Caribbean (LAC). We signal the need to increase adaptive capacities in LAC by going beyond the mainstream risk and climate resilience framework. Social movements, popular solidarity, collective action, local knowledge, experiences on local risk management, and inclusive planning and budgeting are thus considered fundamental for building transformative urban resilience frameworks. We conclude that situated action based on regional priorities is needed, rather than following decontextualised models conceived from welfare societies of the Global North.

Keywords Transformative urban resilience · Uneven vulnerabilities · Social construction of vulnerability · Participatory governance · Urban political ecology

---

P. Ariza-Montobbio (✉)
Latin American Faculty of Social Sciences, FLACSO, Quito, Ecuador
e-mail: pariza@flacso.edu.ec

A. Carrión
Latin American Faculty of Social Sciences, FLACSO, Quito, Ecuador

Instituto de Altos Estudios Nacionales, IAEN, Quito, Ecuador

G. C. Delgado-Ramos
Center for Interdisciplinary Research in the Sciences and Humanities, National Autonomous University of Mexico, UNAM, Mexico City, Mexico

I. Ruiz-Mallén et al. (eds.), *Urban Resilience to the Climate Emergency*, The Urban Book Series, https://doi.org/10.1007/978-3-031-07301-4_5

5.1 Introduction

Climate change impacts are already being experienced in Latin America and the Caribbean (LAC) (IPCC 2014, 2018, 2021; Moreno et al. 2020). Increasing seasonality and intensity of rain, droughts and floods, frequency of extreme weather events, melting and retreat of glaciers, sea-level rise, and other effects of the El Niño Southern Oscillation (ENSO) are affecting urban areas (UN-HABITAT 2011; van Staden 2014; IPCC 2014, 2018; Delgado-Ramos et al. 2020). Climate change risks aggravate existing socio-economical problematics, including human-induced environmental and non-environmental hazards resulting from historical and uneven land uses.

LAC is one of the most urbanised regions globally, with 80% of its population living in cities or metropolitan areas (CEPAL 2017). These urban areas are highly unequal, segregated with 21% of their population, 104.8 million people, living in informal settlements or slums (UN-HABITAT 2016). These marginalised areas are characterised by poverty, overcrowding, poor housing quality, and limited access to basic infrastructures, services, and public spaces; moreover, they are often located in disaster-prone areas and constantly exposed to disease, eviction, and violence (UN-HABITAT 2015;Sandoval and Sarmiento 2020). As Glitz Mayrink et al. (2021) argue, building on Banks et al. (2019), informality is a complex process ridden with power asymmetries involving multiple actors, including those who survive and organise under informality and those who benefit, profit, and accumulate (legally or illegally) by insecure tenure or other social, economic, and political inequalities.

In LAC countries, unequal global exchange shapes socio-economic metabolism (Hornborg 1998; Hornborg 2009). Under global capitalism, the region specialises in capital-intensive *extractivist industries* such as oil and mining and develops urban economies characterised by low value-added services (Eisenmenger et al. 2007). Such dynamics model urban metabolism, plus rapid and expanding urbanisation processes lacking proper planning with a high degree of informality and inefficiencies in land and resource use (Delgado-Ramos 2020, 2021). The increasing consumption of energy and materials is highly unequal: the average urban consumption is more than double the rural counterpart. Such a disparity is even higher between the urban rich and the urban poor. On average, the material consumption of the richest 10% of the urban population is 18 times greater than the poorest 10% (UNEP 2021).

Mitigation and adaptation to climate change and efforts to make human settlements less vulnerable to socio-environmental risks in LAC need to consider these inequalities, which have been deepened further by the COVID-19 crises (CEPAL 2020; Delgado-Ramos and López García 2020). Moreover, the planning for a post-pandemic recovery aimed at a just urban transformation must reconsider mainstream approaches to resilience (Ibid) as they tend to ignore systemic risks and vulnerabilities (Pelling et al. 2021) and leave the burden of city-scale resilience to the individuals or *better* self-organised neighbourhoods, instead of addressing structural inequalities (Meriläinen 2020).

Resilience has been defined by the Intergovernmental Panel for Climate Change (IPCC) as *the capacity of social, economic, and environmental systems to cope with a hazardous event or trend or disturbance, responding or reorganising in ways that maintain their essential function, identity, and structure, while also maintaining the capacity for adaptation, learning, and transformation* (IPCC 2018, p. 557). However, there is a growing diversity of resilience conceptions and narratives (Borie et al. 2019) and a raising call for resilience research and practice to consider power relationships, inequalities, justice, conflicts, and political contestation (Allen et al. 2017a; Eakin et al. 2017; Kaika 2017; Swyngedouw and Ernstson 2019). Far from seeking to maintain the previous status quo, urban resilience aims to move towards more just and equal conditions (Meerow et al. 2016; Delgado-Ramos and Guibrunet 2017). Such a feature may even challenge our own understanding of what is a desirable function of cities[1] and what we can expect from urban life. Due to the reflexive nature of such questions, building or fostering urban resilience in LAC is a highly political issue.

In such a context, this chapter is interested in the key factors affecting urban resilience thinking and action in LAC, by looking at themes, questions, and debates relevant to the region. Through a bibliometric analysis and a qualitative content analysis of 16 interviews to key stakeholders, we address different meanings of urban resilience and identify those common references used to mobilise the concept in contrast or in dialogue with other notions such as urban sustainability. We thus explore what notions of urban resilience are constructed in LAC, and by whom, where, when, how, and why they are promoted.

5.2 Resilience: Perspectives and Critiques

The resilience paradigm and complex systems thinking (Gunderson and Holling 2002) emerged in the context of late and reflexive modernity, in the so-called risk society (Beck 1992), marked by uncertainty, loss of modern institutions legitimacy, and the global ecological crisis (Lampis 2013; Rebotier et al. 2013). Lampis (2013) further argues that resilience is a neoliberal concept. Additionally, in LAC, the recurrent socio-political conflicts, economic crises, and enduring effects of colonialism and neoliberalism, trigger greater uncertainty for addressing urban resilience in the context of rapid and unexpected change. Therein, handling uncertainty is of vital importance and crucial for decision-making (Davoudi 2012; Metzger and Robert 2013).

[1] We refer to the urban function within the prevailing system of (re)production as core spaces of capital accumulation as exposed by Lefebvre (1991), Smith (2010), and Harvey (2008, 2012), but also to the potential function that cities may play in postcapitalist alternatives that prioritise human and non-human well-being based on an urban sustainable, resilient, just, and inclusive urban transformation (Foster 2020; Shingne 2020; UNEP 2021; UNEP and UN-HABITAT 2021).

The resilience paradigm has been expanded to include attempts to prevent and mitigate the disturbances that affect a system (Godschalk 2003). Hence, resilience is commonly associated with the capacity for planning, learning, innovation, creativity, and socio-territorial adaptability (Fleischhauer 2008; Davoudi 2012; Mehmood 2015). Resilience is increasingly gaining attention both in research and public policies design and implementation for disaster risk reduction, climate change adaptation, and more generally, global environmental change (Delgado-Ramos et al. 2020; Egerer et al. 2021; UNEP and UN-HABITAT 2021).

Notwithstanding the success of the concept in capturing the attention of policy-makers, some scholars question its usefulness or relevance both in empirical and theoretical terms (Rebotier et al. 2013; MacKinnon and Derickson 2013). Moreover, the multiple definitions are sometimes imprecise and ambiguous (Lampis 2013). Other authors consider it a *blurred concept* whose application is dependent on the context, scale, and actors involved (Méndez 2012). Others see it as a *boundary object*, having enough plasticity to be adapted to the needs and limitations of different actors that use it but at the same time enough consistency to maintain common ground between actors and places (Borie et al. 2019).

Discussions on the applicability and operability of *urban resilience* are particularly important for regional and urban development. Tentatively, it can be defined as *the ability of an urban system –and all its constituent socio-ecological and socio-technical networks across temporal and spatial scales– to maintain or rapidly return to desired functions in the face of a disturbance, to adapt to change, and to quickly transform systems that limit current or future adaptive capacity* (Meerow et al. 2016, p. 39). Critical analysis of urban resilience requires defining urban systems, their relationship with the territory, the rural or peri-urban hinterland, and the world under globalisation (Filion 2013; MacKinnon and Derickson 2013). Addressing urban resilience also means inquiries about desired urban functions and the purpose of resilience.

The recovery or transformation of the *functions* of a resilient system requires first defining the purpose of the system, whose actors define it or at what time scale and geographical context (Metzger and Robert 2013; Rebotier et al. 2013). How perspectives on urban resilience have travelled, appropriated, and applied is also relevant due to the ideological mobilisation of the concept. In what follows, we briefly describe how urban resilience knowledge is produced and by whom. Far from offering an exhaustive review, we attempt to argue that resilience understanding and practice cannot be taken for granted or equal in all contexts, particularly when the concept's origin has mostly been localised in the Global North.

As Yang et al. (2021) indicate, papers on *resilient cities* indexed in The Web of Science reached 1249 between 2010 and 2019. Nine out of the top ten countries according to publications corresponded to Global North countries which accumulated 73.6% of all publications, being China the only Global South country on that list with 7.8% of all papers analysed; from the LAC region, Brazil barely has a weak presence (Ibid). The absence of publications from the LAC region indicates two issues. First, a weak influence of the LAC academy in the production of peer-reviewed publications in English; and second, most of the urban resilience literature is to be found recently and mostly in Spanish publications, either of scientific nature or grey literature (see below).

In a search of papers dealing with *urban resilience* made in both Scopus and The Web of Science databases, for 1698 publications identified until July 2021, only 95% or 5.6% of them related to LAC case studies. Most of the papers on the LAC region analysed urban climate resilience, resilience of cities to hydrometeorological events (flooding and landslides), the role of nature-based solutions in urban contexts, but also on what respect to earthquake and tsunami resilience of coastal cities (particularly in Chile) and urban security (two papers, one on Mexico and other on Guatemala).

Most of the publications on the LAC region cases are from 2019 (26.3% of the total), 2020 (22.1%) and 2021 (16.8%), which reveals a recent appropriation of the concept in the scientific literature on the region written in English (see Table 5.1). Twelve papers used a LAC case study within the context of comparative analysis. Only one paper did a comparative analysis among two LAC cases; the remaining 11 compared a LAC case with a Global North case. This latter type of paper has usually been framed either from a Global North understanding of urban resilience or from a common shared approach, with limited analysis on how LAC local reality may be explained better from an adjusted or different understanding of urban resilience.

Table 5.1 Urban resilience papers indexed in Scopus and the Web of Science

	Focus of analysis						…of which are comparative analysis with Global North cases
Country of the case studies	Climate-related	Water-related (including flooding and landslides)	Nature-based solutions or similar	Earthquake or Tsunami related	COVID-related	Other	
Argentina	2	2	0	0	0	2	1
Bolivia	0	0	0	0	0	1	1
Brazil	3	5	6	0	2	12	2
Colombia	1	3	1	1	0	1	1
Chile	1	2	1	9	0	4	0
Ecuador	1	0	1	0	0	1	2
Guatemala	0	0	0	0	0	1	0
Jamaica	0	0	0	0	0	1	0
Mexico	10	7	1	2	0	5	4
Peru	2	0	0	0	0	0	1
Puerto Rico	1	0	0	0	0		0
LAC region	1	0	0	0	0	2	0
TOTAL	22	19	10	12	2	30	12
	95						

Source: Authors' own search in Scopus and The Web of Science databases for all years, until July 2021 (search word: *urban resilience* with a sub-search for LAC countries)

In general, most of the papers on the LAC region share a common concern: the importance of informality and poverty as key aspects and even as explicative factors for *chronic shocks* (Fahlberg et al. 2020). In that sense, the need to consider social and political forces has been repetitively emphasised (Eakin et al. 2017), and co-producing resilience by understanding vulnerability-related biophysical and socio-economic, political, and cultural aspects at the local level. Despite such common concern, the literature is mostly theoretical, revealing a gap in planning and building inclusive and participatory urban resilience. Such a gap needs to be closed (Romero-Lankao et al. 2016). For the Global South, and certainly for the LAC region, recognising the relevance of this gap is not enough and becomes imperative bridging science, policy, and practice in a comprehensive manner.

A search of publications in Spanish on *urban resilience* (*resiliencia urbana*) using Mendeley web search tools and Scielo databases detected 90 documents, including articles, postgraduate thesis, workshops, and reports from national and international organisations such as Habitat-Mexico, IDB, among others. However, a search in Google Scholar for the period 2010–2021, suggests 7580 documents, of which 15.8% mention or are focused on Brazil, 13.7% on Mexico and 11.8% on Chile (see Table 5.2). About half of such documents have been published since 2019, which corroborates the recent uptake of *urban resilience* among many academics and practitioners in LAC.

From a sample of 396 documents from Brazil, Mexico, Chile, Colombia, and Argentina, most of the literature in the LAC region deals with climate resilience in cities and metropolitan areas with a perspective that mainly emanates from risk, adaptation to climate change, and urban sustainability (with some contributions that recently linked urban resilience with epidemiological risk due to COVID-19). In

Table 5.2 Distribution of documents on urban resilience by country in the LAC region according to Google Scholar

Country	# of documents	Country	# of documents	Country	# of documents
Antigua y Barbuda	12	Cuba	243	Jamaica	46
Argentina	678	Dominica	11	Mexico	1020
Bahamas	24	Dominican Republic	74	Nicaragua	90
Barbados	23	Ecuador	523	Panama	131
Belize	37	El Salvador	113	Paraguay	82
Bolivia	178	Granada	103	Perú	352
Brazil	1170	Guatemala	141	Puerto Rico	45
Chile	875	Guyana	23	Uruguay	137
Colombia	691	Haiti	54	Venezuela	203
Costa Rica	220	Honduras	108	TOTAL:	7407

Source: Authors' own elaboration. Search in Google Scholar until July 27, 2021. Documents may repeat if they mention or analyse two or more LAC cases

most cases, all these notions are not exclusive, nor are their boundaries, synergies, and trade-offs are clearly distinguished.

Reviewing this literature, we see how urban resilience has been framed in different ways, including those from a socio-ecological and spatial perspective –from which the political economy and the socio-political dimensions of unequal urbanisation are denoted. Therefore, the following aspects are recurrent: informality, inequality/equity, inclusion, quality of life, habitability, material and socio-economic conditions, governance and planning capacities for urban regeneration and transformation, contestation and participatory processes against urban speculation and socio-environmental injustices, and the value of local knowledge and cultural heritage. In that sense, community-based urban resilience is a recurrent issue from which participatory process, co-production of knowledge and the cogeneration of solutions are proposed and even reported. Recurrent issues in such community-based readings are urban participatory planning, social housing, urban services and infrastructures, urban agriculture, and diverse solidarity economy schemes, including house construction, food provision, and services. Other collaborative schemes or networks are interconnected in issues like neighbourhood security, risk prevention and risk management.

Urban resilience as a trending concept in the LAC region has been contested as it is being used for all types of political ideas and discourses. That flexibility or lack of clarity when put into practice, makes the concept weak and, in some cases, even useless for defining and executing local resilience strategies. Its mainstream application tends to be focused on the subject (individual, community, city or socio-ecological system) and its recovery from the *inevitable* crisis instead of on structural conditions influencing the social construction of vulnerability and risk (Lampis 2013). This use of *resilience* into urban practice can, for instance, easily displace responsibility to individual capabilities without taking care of inequalities and power relations that explain uneven vulnerability and resilience between groups, neighbourhoods, communities or cities (Adger 2000; Allen et al. 2017a, b; Lampis 2013; Meerow et al. 2019; Romero-Lankao et al. 2016). Resilience is, therefore, a contested concept, which requires facing questions such as Resilience *for* whom? Resilience *by* whom? To what disturbances? When? Where? How? For what purpose?

5.3 Analytical Framework: Fundamental Questions on Urban Resilience

In this study, we explored key questions, themes, and debates around urban resilience by examining different practices, understandings, and experiences in LAC. Following Meerow et al. (2016), we structured our research around what these authors propose as fundamental questions related to urban resilience. We adapted these questions and incorporated other key relevant queries for the regional

Table 5.3 Fundamental questions related to urban resilience that guided and structured the research in the LAC region

Resilience dimension	Questions
Defining and contextualising resilience	What is urban resilience in the Latin American context? What prominent or relevant initiatives on urban resilience exist and are well known in Latin America?
Who?	Who determines what is desirable for an urban system? Whose resilience is prioritised? Who is included (and excluded) from the urban system?
What?	What perturbations should the urban system be resilient to? What networks and sectors are included in the urban system? Is the focus on generic or specific resilience?
When?	Is the focus on rapid-onset disturbances or slow-onset changes? Is the focus on short-term resilience or long-term resilience? Is the focus on the resilience of present or future generations?
Where?	Where are the spatial boundaries of the urban system? Is the resilience of some areas prioritised over others? Does building resilience in some areas affect resilience elsewhere?
Why?	What is the goal of building urban resilience? What are the underlying motivations for building urban resilience? Is the focus on process or outcome?
How?	Which tools, models or methods can help to build urban resilience in Latin America?
Resilience factors particularly important for Latin America	Are there particular Latin American perspectives on resilience? What elements or region-specific factors influence or shape resilience thinking and action in Latin America? To what extent this concept could be re-signified from Latin America or belongs to a development agenda driven or imposed by the Global North?
Relation of resilience with sustainability	How is resilience different from sustainability? How do these two concepts relate to each other?

Source: Adapted from Meerow et al. (2016)

context (see Table 5.3). Meerow et al. (2016) follow the 5 *Ws*, wondering: *who* determines the *desirable* state of the urban system that should be resilient and for *whom*; to *what* perturbation(s) the urban system seeks to be resilient; and *which* networks or systems are key for such goal as well as the kind of resilience that is leveraged: either to specific threats or to a broader range of variability (i.e. general adaptability). For these authors, understanding urban resilience also requires looking at *when resilience*-building should happen, whether the focus is on short-term disruptions such as storms or long-term stressors such as climate change or long-lasting economic decline. Temporal scales are as important as spatial scales because *where* resilience takes place is influenced by the social construction of urban system boundaries and multi-scale interactions and trade-offs between geographic areas. Actors' underlying motivations, interests and goals are also relevant to inquiry on *why* resilience is promoted in urban areas and certainly on *how* that takes place.

We, therefore, add to the *how* of this frame the relevance and influence of tools, models, and methods used to address urban resilience in the region. We wondered about cross-cutting region-specific factors influencing resilience thinking and action in the LAC context. Finally, we deemed it important to put resilience in dialogue with another mainstream and contested goal: urban sustainability.

These fundamental questions complemented our bibliometric analysis and oriented a set of semi-structured interviews that inform the following sections of this chapter. We consulted experts working mainly in the interdisciplinary fields of urban studies (including urban sustainability), planning, disaster and risk reduction, and climate change adaptation in LAC. Interviewees included academics from research groups (4), practitioners with experience in local urban planning for resilience or with direct participation in local governments associations (4), NGOs (2); activists (2), international cooperation agencies (1), and multilateral agencies (1) (Table 5.4). We conducted group interviews with two to three persons from the same organisation and individual interviews with a representative of the organisation. Their informed consent was obtained prior to each interview, and all interviewees agreed their names to be displayed and published in this chapter. Besides the organisational perspective, we also reported on individual perspectives based on reflective, knowledgeable experiences. With these primary and secondary data, we conducted a qualitative content analysis in search of key themes and debates related to the guiding questions to reflect on urban resilience understanding and practice in the LAC context. When doing our interviews, readings and analysis, we also considered the distinction between *what is* in the current state of affairs and *what should be* in the political realm of desirable alternatives.[2]

The findings described in the following section focus on shared opinions and plausible points of consensus in the conversation about urban resilience in the LAC region. Thus, we do not present or literally quote the individual position of actors, either personal or institutional. We avoid quoting statements as well as the allocation of specific postures unless indicated otherwise.

5.4 Transformative Urban Resilience: Facing Uneven Vulnerabilities in the LAC Region

A first step to contextualise this explorative study on urban resilience is to situate our reflections. The interviews and bibliometric analysis were made after important and simultaneous political protests by diverse social movements in several countries, especially Ecuador, Colombia, and Chile, with a more visible momentum in October 2019, when nationwide multi-actor strikes coincided in the three countries.

---

[2]This dialectical thinking between *what is* and *what should be* is being applied with Marcuse's (1971) *technology as ideology* as a reference framework. A similar and previous application by one of the co-authors can be seen in Ariza-Montobbio and Farrell (2012).

Table 5.4 Actors interviewed

Type of actor: organisation	Geographic area	Interest	Name of persons interviewed	Type of participation
Researcher: Universidade de São Paulo, Instituto de Energia e Ambiente	Brazil, Colombia	Critical approaches to climate change international agenda as uncritically adopted in Latin America. Informal settlements.	Andrea Lampis	In a personal capacity
Researcher: Instituto Internacional para el Medio Ambiente y el Desarrollo (IIED)	Argentina, Colombia, El Salvador	Participation and resilience in intermediate cities, Responding to climate change in cities and in their informal settlements and economies. Community-driven disaster risk reduction and climate change adaptation in urban areas.	Jorgelina Hardoy	In a personal capacity
Researcher: Center for Integrated Mobility Sciences, National Renewable Energy Laboratory.	Boulder, Colorado, USA and Mexico	Climate governance and policy; justice, equity, and inclusion in science, technology, and innovation; environmental justice and equity; urban adaptation and urban resilience in the LAC region.	Patricia Romero-Lankao	In a personal capacity
Research-action:Working group in Latin American Political Ecology of Urbanisation.	Chile, Colombia, Peru	Critiques to resilience as *post-political* and technocratic hiding dissensus and alternative views.	Group interviewed:Francisco Vásquez,Germán Andrés Quimbayo Ruiz, and Sharo López Javier	In a personal capacity
Researcher: Editors of *Urban climates in Latin America* (Henríquez and Romero 2019)	Chile	Direct contact and participation with technical narratives.	Hugo Romero	In a personal capacity

(continued)

Table 5.4 (continued)

Type of actor: organisation	Geographic area	Interest	Name of persons interviewed	Type of participation
Practitioner – Researcher: Resilience Directorate, Metropolitan District of Quito Municipality (MDMQ) and 100 Resilient Cities Initiative (Rockefeller Foundation)	Ecuador	Direct empirical experience on implementing policies, programmes, and projects on urban resilience in Quito and autocritical awareness on the rising power of philanthropic funds.	David Jácome Polit	In a personal capacity
Practitioner – Researcher: Founder of the consultancy firm AAE and academic at Universidad ORT	Uruguay	Experience on urban ecology, nature based-solutions, urban planning, urban sustainable transformation.	Diego Martino	In a personal capacity
Practitioner – Local Governments Associations: ICLEI (International Council for Local Environmental Initiatives- Local Governments for Sustainability)	Brazil	Experience on local urban planning for resilience.	Rodrigo Corradi	In a personal capacity
Practitioners: General Directorate of Resilience, Secretariat of Integrated Risk Management and Civil Protection of Mexico City	Mexico	Experience on urban resilience, risk management and civil protection.	Group Interviewed: Norlang Marcel García Arróliga, Martín Alejandro Guerra Ramírez, and Aideé Stephanie Jiménez Ávila	Institutional position
Practitioner: Former director of Mexico City's Resilience Agency and co-founder of Ithaca Environmental consultancy	Mexico	Experience on urban resilience planning and implementation as well as in urban socio-environmental challenges associated with global change.	Arnoldo Matus Kramer	In a personal capacity

(continued)

Table 5.4 (continued)

Type of actor: organisation	Geographic area	Interest	Name of persons interviewed	Type of participation
NGO: Habitat for Humanity	Central American Dry Corridor	Experience on resilience initiative for small Caribbean states from a humanitarian aid NGOs perspective.	Group Interviewed: Mario Roa Romero and Zulma Espinoza	Institutional position
NGO: *Ciudades Resilientes al Clima (CRC)* initiative. Fundación Futuro Latinoamericano (FFLA) y Alianza Clima y Desarrollo (Climate and Development Network, CDKN)	Regional, Ecuador, Argentina	Experience on projects and initiatives on urban resilience in Latin American from the perspective of NGOs involved in climate action.	Group interviewed: María José Pacha and Gabriela Villamarín	In a personal capacity
Activist: Observatorio Latinoamericano de Acción Climática (OLAC) - GFLAC (Climate Finance Group of Latin America and the Caribbean)	Regional perspective	Climate finance in Latin America.	Carola Mejía	In a personal capacity
Activist: ActionLac (Climate Action Network in Latin America)	Chile	Critical perspectives on citizen action for climate change.	Karla Maas	In a personal capacity
International cooperation agency: GIZ	Mexico	Experience on urban resilience planning, research and implementation.	Auribel Villa	In a personal capacity
Multilateral agency: CEPAL, Human settlements division	Regional perspective	View from staff of multilateral organisations.	Diego Aulestia	In a personal capacity

Source: Authors' own elaboration

Months later, the COVID-19 outbreak led to deeper health, social, political, and economic crises worsening the already existing inequalities and vulnerabilities, particularly but not only in urban contexts where most of the cases have been confirmed (Delgado-Ramos and López García 2020; Filgueira et al. 2020; Pelling et al. 2021). Migration from the Caribbean and Central America to Mexico and the United States, but also between Venezuela and Colombia, has aggravated the humanitarian crises but also the highly complex political and socioeconomic context in which such phenomenon takes place (particularly in transborder cities) (Cantor 2018; LAWG 2020; Selee et al. 2021).

These circumstances influenced any debate on resilience and, of course, our research as well as our interviewees' answers. The socio-political consequences of both key events, *social protests in October 2019* and *COVID-19,* influenced discussions on *transformative reactivation* or *transformative recovery* campaigns both from institutions, such as CEPAL, and social movements or activist groups in the region, such as Latin American Observatory for Climate Action (OLAC-GFLAC), Climate Action Network in Latin America (ActionLac) or Climate and Development Network (CDKN). Consequently, the agendas on *urban transformation* proposed by UN-HABITAT, UNEP or international/regional finance institutions, NGOs, cooperation agencies and academia have also been affected (WCRP 2019; Agarwala and Vaidya 2020; IRP 2020; OECD 2020; UNEP and UN-HABITAT 2021; UN-HABITAT 2021). In Chile, despite the difficulties created by the pandemic, a constitutional process is ongoing (as of 2021). The type and timing of recovery plans set in motion in the following years will certainly define investments and actions for promoting business as usual or a transformational change; including urban sustainability and resilience (Hepburn et al. 2020; Solecki et al. 2021). Likewise, daily evidence of present and potential future scenarios of climate change effects and the uneven distribution of its impacts (Guivarch and Taconet 2020; Gazzotti et al. 2021) also contributes to situating the debate, especially on structural and socially constructed inequalities and vulnerabilities, as the on-set for urban resilience.

5.4.1 What Does Urban Resilience Mean in the Latin American Context?

Departing from current official, *mainstream* or *status quo* definitions, our interviewees framed the notion of urban resilience into more mechanical views focused on the recovering capacity of critical urban functions, networks, infrastructures, and services. All interviewees agreed that resilience is a concept defined and constructed in the Global North, while most circulating definitions in LAC do not explicitly incorporate a contextualised view that recognises the differences of what urban resilience may mean in the region (or even in the Global South in general). They also agreed on the limited region's agency, capacity, and knowledge on the topic despite the existence of valuable work, particularly around risk assessment, management and even prevention (which is less usual). Such limitations occur in the face of vibrant contestation processes related to urban political economy, urban political ecology dynamics and forced displacement in the region's cities (see for example Cardona 2004; Cardona et al. 2010; García Acosta et al. 2012; Cuvi 2015; Zeiderman 2016; Few et al. 2021). Moreover, some interviewees explicitly mentioned that resilience is broadly applied in LAC with either a predominant biophysical view focused on the technological recovery of infrastructures or with a neoliberal view focused on individual adaptive capacities rather than collective, societal, and political power structures. These tendencies, along with the lack of recognition of endogen knowledge and agency, have sometimes reinforced the disconnection, first, among risk and climate change thinking and action, and later, between climate change and resilience perspectives.

By looking to a normative vision on *what urban resilience should be* –at least from a Latin American viewpoint–, a more systemic and transformative approach came to the front, most interviewees argued. Transforming inequalities and social and power structures that historically have built uneven vulnerabilities are perceived as central for allowing any meaningful practise of urban resilience.

Securing food sovereignty, public health, water, sanitation, among other basic urban services, is a precondition for an urban resilience notion that puts equity and justice at the forefront. If people's rights are not fulfilled and addressed, but also if nature's functions and biodiversity are not well preserved (Oke et al. 2021), there is no room for urban resilience because there would not be a *desirable* state. That is why interviewees, in general, understood that urban resilience should be transformative, promoting the recovery from crises, disasters or extreme events in a way that addresses the changes needed in existing, previous, and potential problems that may reproduce vulnerability, multi-dimensional poverty, gender and other inequalities, and ecological rift (Foster 2011; see also Chap. 2 to look further into transformative approaches). This general agreement among interviewees on a transformative approach to resilience is one key finding of our research. It reveals that addressing (pre)existing structural problems and the power structures that propelled them, alongside individual and collective agency capabilities that shape urban governance, are departure points for any discussion of urban resilience in LAC.

5.4.2 Common References to Resilience Programmes and Initiatives

When asking the interviewees for common references or models of urban resilience initiatives, two emerged as the most visible and quoted: the *100 Resilient Cities* by the Rockefeller Foundation and the *C40* programmes undertaken mainly in large cities or metropolitan areas. Other references were the United Nations Office for Disaster Risk Reduction (UNDRR), Making Cities Resilient 2030 (MCR2030), and the related Ten Essentials of the Sendai Framework for Disaster Risk Reduction. UNDRR approaches were intended to integrate specific adaptation to eventual risks but seem still under construction. Reactive and palliative interventions centre the attention, instead of more integral and structural action to prevent disasters and reduce risk by addressing inequalities and warranting access to good public services and dignified homes for all.

Other common references were documents, projects, and programmes from local government associations such as ICLEI Latin America (International Council for Local Environmental Initiatives), RAMCC (*Red Argentina de Municipios frente al Cambio Climático*) or Adapt Chile (for a general review from an adaptation perspective, see Delgado-Ramos et al. 2020). ICLEI, RAMCC and Adapt Chile are of particular interest due to several guides for local climate action they have published and diffused (see ICLEI 2016; Mitchell et al. 2019).

Next, we first present our findings on the *who, what, when, where, how* and *why* on urban resilience in Latin American and its relation to urban sustainability, followed by an empirical analysis of the 100 Resilient Cities programme.

5.4.3 Resilience for Whom? Who Decides and Participates?

The illegal and insecure land tenure of inhabitants living in risk-prone and marginalised urban areas tend to exclude these populations from decision-making processes (Rivero-Villar and Vieyra Medrano 2021). According to most interviewees, these collectives are usually unrecognised, both legally and politically, as active participants and, furthermore, as potential decision-makers. Recurrently, experts and politicians lean to oversee or underestimate the effects of population relocation or exclude poor neighbourhoods following a cost-effective perspective. Urban authorities do not consider low-income neighbourhoods as a source of taxes, revenues, and economic growth. Further, decision-makers perceive the private corporate sector as the most secure and economically efficient pathway to promote resilience.

According to interviewees, participatory planning and decision-making are desirable, especially alongside vulnerable and poor communities living in informal settlements located in risk-prone areas. This participation should be under co-management schemes, going beyond consulting or coordinating exercises where technical experts and politicians take the most critical decisions. Action-research projects such as *cLIMA sin riesgo* in Lima, Peru, are steps towards this desirable participatory governance (Allen et al. 2017b).

5.4.4 Resilience to What?

The focus and priorities of urban resilience are context-specific, dependent on the geographic scale (*where,* see below), the actors involved (*who*, see above) and the desired functions and purposes of the urban systems which are decided to preserve or change (*why*, see below). That is why, when asked for the most important perturbations that need to be addressed or the networks or systems to protect, most interviewees referred that the focus should be on assessing uneven vulnerabilities and the social processes and power relations that construct them. Ensuring basic services and attending human rights, but also urban and non-urban ecosystems' resilience should be the focus of urban resilience. While floods, droughts and other climate change impacts are relevant for urban resilience, interviewees consider that pre-existing inequalities influencing human responses to them are of prior importance.

Geographic contexts and ecosystems along LAC would face distinct climate change impacts. Some areas will be notably more affected by floods or landslides, while heatwaves, droughts or earthquakes will impact others more. Cities of different sizes and functional roles would have different knowledge and economic resources available for mitigation and adaptation. Capital cities and metropolitan regions with a central economic and administrative role would likely attract more resources and a broader capacity for action. In contrast, intermediate cities as nodes of regional commerce and rural-urban mediation would feasibly address key local problems. However, interviewees perceived that private corporate actors and elites would be centred on recovering economic growth, whereas marginalised and impoverished workers would be focused on assuring dignified living conditions. Thus, instead of thinking on specific adaptations to individual threats, most informants agreed that, in the LAC region, the focus should be on general adaptability. Building adaptive capacity through education on systems thinking, for example, could create enabling conditions to transform urban planning. In this case, capacity building refers not only to institutional capacities but also to the inclusion of grass roots organisations and lay individuals, allowing cross-fertilisation of ideas and co-production of solutions as robust as possible.

In this context, the inclusion of local knowledge and practices in urban resilience decision-making is desirable. Many ecosystems in LAC, such as the Andean mountains, are marked by daily and seasonal climate variability. Living under uncertainty is precisely part of indigenous knowledge that has historically been ignored in decision-making processes. Socially constructed vulnerability and unpredictable ecosystem variability have forged everyday climate resilience experiences by excluded populations, whose knowledge could be of great value to change broader societal trajectories.

5.4.5 When to Act for Resilience?

In LAC, disaster risk reduction has been implemented mainly from a reactive perspective, privileging ex-post interventions rather than preventive actions and structural changes, which can reduce exposure and sensibility of populations living at risk. Short-term interests usually lead decision-makers to prefer *business as usual* scenarios that reproduce decontextualised temporary solutions. Thus, mainstream resilience conceptions tend to privilege resistance to or recovery from rapid on-set disturbances rather than transitions or transformations to face slow-onset changes. According to some interviewees, when the focus was on reactive action to disasters rather than structural transformations, resilience practice was perceived as following obsolete and maladapted practices for rebuilding urban habitats. In many LAC cities, relocating populations in standardised new urban developments by moving them away from their previous affected homes could derive on important socio-ecological consequences, eroding social networks, livelihood systems and even local ecosystems (Wolff 2020). Thus, reacting to more frequent and intense extreme weather events seems to be less resilient in the long run than adopting a gradual transition approach directed to reconcile short-term actions with incremental long-term transformations. The interviewees also argued that concrete local actions could further mobilise and engage the public by raising the necessary awareness to confront the deep societal changes needed to build better futures for all, particularly for the most vulnerable.

5.4.6 Where to Act for Resilience?

The experience of practitioners interviewed reveals that urban sprawl complicates the definition of spatial boundaries, especially since most of perturbations or challenges transcend the arbitrariness of administrative limits, such as in metropolitan and megalopolitan regions. Deregulated suburbanisation complicates urban planning for resilience. Any intention to act on informal settlements gets trapped into a paradox: marginalisation secludes the most vulnerable people in risk-prone areas without legal tenure and local governments do not recognise such areas for further intervention. Lack of formal recognition refrains resilience in-so-far urban planning is legally bound to focus on consolidated areas; also, investing public resources in the middle to high-income areas could have higher returns (including political gains). In LAC cities, urban resilience needs to focus on reverting the uneven distribution of vulnerabilities, by recognising that the current urban landscape relies on displacing unwanted land uses or environmental *bads*, such as air and water pollution, to poor neighbourhoods (Romero et al. 2010; Mendes et al. 2020). Social segregation and uneven distribution of environmental *goods* and *bads* lead to a closer interrelation between *who* should be the main actors of resilience and *where* the priorities of action should focus.

Urban resilience projects should be aware of geographical trade-offs and their indirect effects. For example, green infrastructure could foster resilience with the risk of gentrification and displacement (Anguelovski et al. 2019; Rigolon et al. 2018). Relocating populations could improve some areas at the expense of others. Property insurance needs to consider inequalities by distributing costs between higher-income households (e.g. via property tax), the private sector and/or the public sector. This latter measure already operates in Manizales (Colombia), where a collective insurance policy has been implemented since 1999. The *Manizales predio seguro* governmental programme redistributes voluntary insurance of high-income properties by extending coverage to low-income properties at no extra cost. The insurance is paid via the property tax, exempted to low-income properties (ERN n.d.). There are also early and very limited experiences in Querétaro (Mexico), where the government acquired an insurance policy since 2020 to cover losses due to house and business robbery, flooding, or fire (limited to a very low deductible amount that ranges from 2500 and 10,000 US dollars in the event of a claim). The insurance is granted to those properties whose owners have paid their property taxes during the first month of the fiscal year.[3]

5.4.7 How to Build Resilience?

Inquiring *how* to build urban resilience refers to methods, tools or models for urban resilience which could interrelate the previous questions by proposing forms of decision-making and prioritisation. Such methods should allow for a participative *who* and a transparent *what, when* and *where*, considering mid to long-term thinking and anticipating potential trade-offs. According to some interviewees, lack of good quality data at appropriate scales and resolution, together with weak institutions troubled by discontinuity of both qualified personnel and robust policies, constrain the incorporation of resilience in decision-making processes and implementation actions. Recurrently, government changes and limited human capital tend to halt policy success and coordinated planning.

As mentioned by interviewees, promoting a *systems thinking perspective* through educational actions and targeting the general public, technicians and urban planners is a first step to address such problems and advance towards more integrated and coordinated planning and decision-making. Multi-stakeholder platforms and hybrid citizen-expert forums, together with multi-scale vertical and horizontal institutional coordination, are tools for enhancing participatory planning, which is also in line with current urban literature (WCRP 2019; Solecki et al. 2021). Democratising decision-making, getting closer to peoples' languages and local knowledges to better communicate the challenges of climate change, and taking situated decisions are urgent measures to face the democratic flaws of incomplete citizenry and human rights violations.

A systemic approach oriented to address urban nexuses, such as water-energy-food and implementation plans that integrate watersheds, bio-regional or ecological units beyond administrative units, can be a way to implement an integral perspective of urban resilience. Such assessment, noted by interviewees, could help manage trade-offs and overcome limiting rural-urban dichotomies and administrative conflicts over competencies between sectors or institutions. In several LAC countries, climate-environmental problems are not always integrated into planning instruments or operate in silos, with limited capacities at the local level (Delgado-Ramos 2021; UNEP 2021). The absence of robust institutional or metropolitan coordination is also common, with very few and still limited exceptions. The Megalopolis Environmental Commission (CAME), legally in charge of coordinating climate and environmental actions in 240 municipalities of six central states in Mexico (including the metropolitan area of Mexico City), is an exemplary case. However, its incidence has mostly focused on air quality, with limited or very limited actions on

[3] https://seguropredialqueretaro.com

biodiversity conservation, water management and climate change which are also its attributions. In addition, the coordination of the CAME with Civil Protection is limited, and therefore the possibility of moving forward to a coordinated megalopolitan resilience agenda is still constrained.

Proposals for collaborative governance and planning based on a systemic approach call for a gradual transition towards more plural, inclusive, and participatory decision-making. Furthermore, consolidating collaborative governance and planning strategies could lead to a more fundamental transformation towards a radical democracy.[4] Sustained and encompassing participation processes could help to build systemic, preventive, and integral responses rather than punctual reactions to stressors. By strengthening democracy in such a way, discussions and decisions can be open about why to act for resilience and for what societal purposes.

5.4.8 Why Is Resilience Needed? Rethinking the Meaning and Purpose of Resilience

We can see the difference between the goals of two different resilience approaches. The first one guarantees access to basic services and urban infrastructures that support urban functions, life quality and dignified homes for all. The other prioritises capital accumulation by assuring minimum conditions for the working class, improving infrastructures for the elites and productive processes that do not always account for environmental externalities.

Resilience has also become a label for any urban project to be sponsored by international financing institutions, not all actually of a transformative nature. As noted by some interviewees, while climate change is part of the priorities of new international financing schemes, in practice, many *climate* projects tend to address wicked recurrent development problems, such as basic services or dignified homes.

Developing adaptive capacities and building robust participatory processes for urban planning and democratic decision-making appears to be key for making a gradual transition towards urban resilience. Forging a common vision between informal and marginalised sectors and sensitive planners to guide and contextualise the meaning of daily activities within broader processes seems fundamental. Interviewees remarked the need for combining, in parallel, structural transformations with the necessary responses to recurrent disasters and risks (such as epidemiological risks).

The transformation of structural inequalities and power relations towards democratic fulfilment of human rights and food sovereignty needs to be addressed before, during and after any disturbance, with the aim to transform the underlying social

[4] We refer to a broad notion of radical democracy, which could include different perspectives, agonistic, deliberative or autonomist, all seeking to expand liberal definitions of democracy, including differences to equality and freedom (Little and Lloyd 2009).

causes of vulnerability. Additionally, although not directly addressed by interviewees, a non-anthropocentric approach is certainly needed to support such transformation, and it could be promoted through the advancement of the so-called *more-than-human cities* or *multispecies cities* (Houston et al. 2018; Robertson 2018; Connolly 2020).

5.4.9 Resilience and Sustainability: Synergies and Contradictions

Resilience and sustainability are seen as complementary. Sustainability focuses on long-term duration, intergenerational equity, and the equilibrium between social, economic, cultural, and ecological dimensions. The sustainability discourse arouses four decades ago out of growing evidence of the environmental impact of advanced and globalised industrial societies. Resilience focuses on responding to any crisis and emerges due to the growing complexity and uncertainty of contemporary climate-environmental problems. Resilience thinking assumes perturbation and instability as inevitable and internal to the very functioning of (urban) socio-ecological systems. Sustainability contributes to resilience by reducing environmental impacts, assuring resource conservation, and promoting the long-term viability of livelihoods. Resilience contributes to sustainability by fostering adaptive capacity and flexibility, allowing better preparedness for disruptions. However, while interviewees reported these synergies between both concepts, they also saw contradictions and trade-offs. For instance, building resilience often implies redundancy in infrastructures, reducing efficiency and increasing monetary and energy costs, likely to increase consumption and emissions (Delgado-Ramos and Guibrunet 2017; UNEP 2021).

The *naturalisation* of crises, implicit to the resilience notion, could also depoliticise the discussion by distracting the attention away from the social construction of disasters and the induced crisis for capital accumulation, which lie at the heart of the unsustainability of industrial capitalism and the so-called neoliberal city (Rossi and Vanolo 2015). Recognising the social construction of crisis and disasters and their relation to economic development and social inequalities is fundamental to promote synergies and address contradictions between resilience and sustainability, aiming to inform policymaking in LAC. Under globalisation, this is even more important given the imposed economic role to the region as an exporter of cheap raw material and importer of expensive manufactured goods (Samaniego et al. 2017; Svampa 2019).

5.4.10 The Initiative of 100 Resilient Cities in the LAC Region

In what follows, we explore further the case of the 100 Resilient Cities (100 RC) initiative to stand out key aspects and challenges for building urban resilience in the LAC region. In doing this, we offer a more empirical reading of our previous analysis.

Based on systems thinking and general adaptability, these 100 experiences endeavoured institutionalising and implementing local resilience strategies rather than specific adaptations to individual hazards.[5] The initiatives had mixed outcomes: on the one hand, due to a diverse understanding of resilience (the *what* and *why* dimensions); on the other, because of the diversity of leadership skills, local socio-political contexts, and dynamics, as well as to regulatory and institutional arrangements prevailing in each host city. In LAC, 16 cities participated in the initiative, mainly metropolitan and capital cities.[6]

Despite the recurrent recognition of informality, marginalisation, and poverty as key aspects in building urban resilience in LAC, a notorious constraint of 100 RC's resilient strategies was the enunciative nature of such aspects. Hence, there was insufficient analysis and understanding on *how* to operationalise urban resilience building, particularly within informality (which is not exclusively related to urban poverty and marginalisation). Most implementation strategies limited their attention to state the challenge that informal settlements and informal services represent, with a constrained consideration of the relationship between the who and the how dimensions of resilience. Consequently, none of them presented a comprehensive and robust understanding of urban resilience building in the context of informality, nor proposed implementation mechanisms for coordinated actions within informal areas.[7]

---

[5] The general appreciation is, however, positive. Once the Rockefeller-Arup initiative was formally closed, the emergence of two novel initiatives probed the *resilience* of the original enterprise. On the one hand, the Resilient Cities Catalyst, an independent nonprofit organisation was created. On the other, the Global Resilient Cities Network, replaced the original 100 cities' network. The later launched more recently a coalition for urban recovery from COVID-19, called Cities for a Resilient Recovery – C2R. The 100 RC initiative also validated the value of cross-fertilisation and replicability of experiences, certainly not without challenges and limitations as described in this section.

[6] Cities participating in the initiative were: Buenos Aires, Cali, Ciudad Juarez, Colima, Guadalajara, Medellin, Mexico City, Montevideo, Panama, Porto Alegre, Quito, Rio de Janeiro, Salvador, Santa Fe, Santiago de Caballeros, and Santiago.

[7] In the best of cases, scattered actions dominated, being among those mentioned the following: improving informal settlement habitat areas and requalifying informal settlements as actions of an integration strategy, incorporating informal waste management into conventional or circular waste management schemes and even one case seeking to eradicate informal activities in landfills, ordering informal transport, inclusion of informal volunteer organisations to orderly contribute to disaster response, mapping formal and informal economic activities for promoting cultural and creative economic schemes, and one case aiming to reduce informal vendors in public spaces. Informality beyond the poor and marginalised were mostly not recognised. The cases of Rio de Janeiro and Santiago (Chile) denote the total absence of informality considerations.

The above-mentioned deficiency relates to an ongoing learning curve,[8] a process modelled by LAC cities' diversity and complexity of political, legal, economic, social, and cultural features. This certainly has been poorly analysed – an aspect that aggravates for small cities which have not been part of the 100 RC initiative.

In small and mid-size cities, where more than half of LAC urban population lives, one interviewee considered two possible situations, both characterised by weak institutional capacities and comprehension on *how* to put into practice local resilience building and how to prioritise actions, particularly concerning the *where* and *when* of resilience. On the one hand, an absence of strategies and actions explicitly recognised as part of a resilience agenda. On the other, a set of scattered actions that unintentionally or indirectly support resilience building and that not necessarily may address informality and potential trade-offs with other local action agendas. Therefore, there is an urgent need for a better understanding of *what* resilience means in practice for particular contexts. Enhancing local capacities and enabling institutional changes is also needed.

Another challenging issue that 100 RCs confronted in LAC emanated from the assumption of the existence of relatively solid institutional frameworks to support the project, which are actually difficult to find at the local level. Furthermore, some 100 RC experiences faced the challenge to achieve full acceptance within current political and power institutional structures, sometimes ignoring created interests and attempting prevailing institutional attributions. They also faced policy volatility and other political instability contexts. These last elements seem to be the main cause of the inability of some cities to develop a resilient strategy document,[9] or the incapacity of others to plan and put into motion a resilience strategy working close to the city's mayor (something that was expected from the 100 RC initiative and that actually happened in most of the cases). All these elements had implications in the *how* of resilience building.

The 100 RC initiative advocated for the co-production of knowledge and solutions. Yet, since each city relied on local capacities, the participatory processes and knowledge produced were of diverse nature and did not always include a direct interaction with the most vulnerable communities at earlier design stages. The flexibility of the methodology led to relatively robust community-based participatory experiences such as the one in Medellin (Colombia), but others focused on the inputs offered by NGOs and organised civil society, such as the strategy of Mexico City.

The case of Mexico City exemplifies the challenges involved in delineating a strategy for a megacity, which requires a broad vision and coordination while simultaneously involving participation at the neighbourhood level. Such experience denotes that not only *who* participates and *how* participation is shaped, but also

[8] We refer here to the process of becoming proficient on the work of building resilience in relation to the experience available. Given that the experience is still limited, the proficiency is still ongoing.

[9] It has been reported that 77 out of 100 cities developed a resilience strategy (www.rcc.city/our-history).

where and *when* it takes place, are all aspects with important implications on building urban resilience.

An insufficient understanding of communities' priorities and expertise could lead to their weak engagement in the initial stages. An approach mostly dominated by experts may not necessarily place community-based practices at the forehead of the process for delineating and prioritising urban resilience strategies. As a result, conventional businesses schemes for supporting economic growth may for instance dominate, limiting or excluding ongoing popular and solidarity economy relevant in building urban resilience (i.e. informal circular and solidary economic channels for food supply, reuse of goods and recovery of materials, to informal schemes of peri-urban water supply, waste management and even transportation).

In LAC, the 100 RC experiences also reveal coordination challenges. The cross-sectoral and interdisciplinary nature of resilience requires a greater degree of intra- and inter-institutional coordination between different agencies and sectors, yet the dominant approach in urban planning and management in LAC is mostly based on siled structures. Creative schemes are thus needed to explore collaboration pathways and alliances, a challenge that applies not only to cities (administratively defined) but also to broader urban systems, including metropolitan, megalopolitan, and transboundary urban regions. In the latter case, limited communication seems to be usual, as well as the difficulty for coordinating transboundary actions due to international legal agreements.

For example, El Paso, Texas, and Ciudad Juarez, Chihuahua, are both part of the 100 RC initiative and conform a continuous urban environment that shares vulnerabilities and dynamics in terms of biophysical risks, complex circulation of resources, goods and people, and unequal urban space production practices and management (Guneralp and Delgado-Ramos 2018). Current transborder regulations, however, restrict binational cooperation in the case of emergencies, a reality that has led to some local policymakers to highlight the need of working on innovative mechanisms that may resolve these constraints, starting with a fluid communication among decision-makers, for example, thru the creation of a binational *emergency room* but also by strengthening the communication and collaboration among resilience officers in El Paso and Ciudad Juarez. Moreover, a process for building stronger capacities in local decision makers, but also in mass media, is expected to start in 2022 with the support of scientists and media associations and networks.

Finally, as in any other experience, success may be affected, positively or negatively, by leaderships, political changes, capacity building, social contestation, and certainly by COVID-19 impacts. The latter may become an opportunity to accelerate urban resilience planning and building, fully considering the interaction of the who, what, when, where, why and how dimensions within a context of noticeable informality, inequality and limited community-based participation at earlier stages.

5.5 Conclusions

Climate change impacts are being experienced in urban areas of LAC. They are unequally distributed among social groups, ecosystems, and neighbourhoods due to structural and socially constructed vulnerabilities. In the LAC region, socio-economic and urban metabolism are influenced by unequal global exchange and rapid, deregulated, and unplanned urbanisation, characterised by unequal land and resource consumption patterns.

The prevalent lack of economic diversification and consolidated economic structures affects alternatives to development and limits the adoption of urban resilience policies. Extractive industries and low-value-added services, frequently informal, contribute to a regional and urban development characterised by inefficient linear urban metabolic patterns, which are synergistically reinforced by urban forms and land uses that promote land mono-uses, urban sprawl and overcrowding. These factors constitute an unequal production of urban space, where diffuse suburbanisation of informal settlements and *gated* communities combine with formal and dense neighbourhoods (Abramo 2012; Broto et al. 2012; Coq-Huelva and Asián-Chaves 2019). Inequalities, poverty, and violence, especially against women and migrants, together with weak human rights enforcement, compromise access to basic services and human rights fulfilment, a clear limitation for resilience. These structural socio-economic and socio-political factors constitute key issues for meaningful urban sustainability and resilience in the LAC region. However, mainstream urban resilience policies are limited in their understanding of urban informality and inequalities and address them partially or insufficiently. Informal settlements are often directly or indirectly excluded from decision making or tend to be included in later stages of participation. Context-specific vulnerabilities are often overlooked by standardised *business as usual* measures such as relocation or land tenure regularisation schemes.

Valuing LAC specific characteristics, situating and contextualising urban resilience initiatives requires a *systems thinking perspective* which integrates and interrelates a participatory and inclusive *who* with a transparent *what, where* and *when,* which consider long-term vision and try to anticipate potential trade-offs. Democratic multi-actor and multi-scalar decision-making (*how*) should recognise political discussions and differences and address multiple and contested purposes and visions (*why*). Deciding and acting for urban resilience requires a sensitive approach to trade-offs and inequalities between social groups, ecosystems, and geographic areas, as well as managing dynamic equilibriums between societal goals such as sustainability, equity, and resilience.

Ecosystem diversity becomes an opportunity for integration of traditional knowledge and modern science for agroecological, ecological restoration and nature-based solutions oriented towards popular and solidarity economy, rather than reproducing uneven urban development based on grey infrastructure. Popular solidarity and collective action to confront crises and disasters have been recurrent and key for successful recovery in past experiences in LAC. Local government participatory planning and budgeting, together with social movements and neighbourhood organisations, are also important available resources for building resilience (Solecki et al. 2021).

A transformative and more-than-human approach to urban resilience should work towards changing structural power relations, combining short-term actions with long-term transitions, a process that requires transcending colonial approaches and technocratic resistances. For example, by framing and addressing resilience within the diversity and complexity that characterise the political, economic, social, and cultural reality of the region. It also presupposes challenging the idea of continued economic growth and instead centring our attention on human and nature prosperity and flourishing in both urban and rural settlements. In that sense, strategies based on decoupling (delinking economic growth from resource and land use) and circular economy models (efficient use of resource and reduction of waste) have to be criticised and contested because, on the one hand, they tend to preserve and even reinforce current political-economic and geopolitical relations and thus inequality, and on the other, because the applicability of such strategies under capitalist relationships is limited and insufficient to generate the transformation required to move towards just sustainable and resilient cities for people and nature.

Acknowledgements This work was supported by the International Development Research Center of Canada, under Grant N° 108443-001, as part of the project 'Building leadership in LAC cities in a changing climate' implemented by Ariza-Montobbio and Carrion at the Latin American Faculty of Social Sciences (FLACSO Ecuador). It is also an outcome of the Texas A&M/CEIICH-UNAM-CONACYT project 'Food-Energy-Water Security in a Transborder Urban Context', under the Award N° 2018-014.1, developed by Delgado-Ramos from the Center for Interdisciplinary Research in the Sciences and Humanities, and Burak Guneralp from the College of Geosciences of Texas A&M University.

References

Abramo P (2012) La ciudad com-fusa mercado y producción de la estructura urbana en las grandes metrópolis latinoamericanas. EURE, revista latinoamericana de estudios urbano regionales 114:35–69

Adger WN (2000) Social and ecological resilience: are they related? Prog Hum Geogr 24(3):347–364. https://doi.org/10.1191/030913200701540465

Agarwala P, Vaidya H (2020) Can COVID-19 fill the void of City Governance for Urban Transformation? UN-HABITAT

Allen A, Griffin L, Johnson C (eds) (2017a) Environmental justice and urban resilience in the global south, Political science series. Palgrave Macmillan, New York

Allen A, Belkow T, Estrada CE, de los Ríos S, Kamiya M, Lambert R et al (2017b) De la mitigación de desastres a la interrupción de trampas de riesgo: La experiencia de aprendizaje-acción de cLIMA sin Riesgo. REDER 1(1):6–28. http://www.revistareder.com/ojs/index.php/reder/article/view/2

Anguelovski I, Irazábal-Zurita C, Connolly JJT (2019) Grabbed urban landscapes: socio-spatial tensions in green infrastructure planning in Medellín. Int J Urban Reg Res 43(1):133–156. https://doi.org/10.1111/1468-2427.12725

Ariza-Montobbio F, Farrell KN (2012) Wind farm siting and protected areas in catalonia: planning alternatives or reproducing one-dimensional thinking? Sustainability 4(12):3180–3205

Banks N, Lombard M, Mitlin D (2019) Urban informality as a site of critical analysis. J Dev Stud 56(2):223–238. https://doi.org/10.1080/00220388.2019.1577384
Beck U (1992) Risk society towards a new modernity. University of Munich, Germany, Munich
Borie M, Pelling M, Ziervogel G, Hyams K (2019) Mapping narratives of urban resilience in the global south. Glob Environ Chang 54:203–213. https://doi.org/10.1016/j.gloenvcha.2019.01.001
Broto VC, Allen A, Rapoport E (2012) Interdisciplinary perspectives on urban metabolism. J Ind Ecol 16(6):851–861. https://doi.org/10.1111/j.1530-9290.2012.00556.x
Cantor DJ (2018) Cross-border displacement, climate change and disasters: Latin America and the Caribbean. UNHCR
Cardona OD (2004) The need for rethinking the concepts of vulnerability and risk from a holistic perspective: a necesssary review and criticism for effective risk management. In: Brauch HG, Oswald Spring U, Mesjasz C, Grin J, Kamerl-Mbote P, Chourou B, Dunay P, Birkmann J (eds) Mapping vulnerability: disasters, development and people. Earthscan, London
Cardona OD, Bertoni JC, Gibbs T, Hermelin M, Lavell A (2010) Understanding and managing risk associated with natural hazards: an integrated scientific approach in Latin America and the Caribbean. In: Science for a better life; developing regional scientific programs in priority areas for Latin America and the Caribbean, vol 2. ICSULAC-CONACYT, Rio de Janeiro/Mexico City
CEPAL (2017) Panorama del desarrollo territorial en América Latina y el Caribe, 2017: agendas globales de desarrollo y planificación multinivel. Comisión Económica para América Latina y el Caribe (CEPAL), ILPES, LC/TS.2017/120, Santiago de Chile
CEPAL (2020) Construir un nuevo futuro: una recuperación transformadora con igualdad y sostenibilidad. Comisión Económica para América Latina y el Caribe (CEPAL), Santiago
Connolly C (2020) From resilience to multi-species flourishing: (Re)imagining urban-environmental governance in Penang, Malaysia. Urban Stud 57(7):1485–1501. https://doi.org/10.1177/0042098018807573
Coq-Huelva D, Asián-Chaves R (2019) Urban sprawl and sustainable urban policies. A review of the cases of Lima, Mexico City and Santiago de Chile. Sustainability 11(20):5835. https://doi.org/10.3390/su11205835
Cuvi N (2015) Un análisis de la resiliencia en Quito, 1980–2015. Bitácora 25(2):35–42
Davoudi S (2012) Resilience: a bridging concept or a dead end? Plan Theory Pract 13(2):307. https://doi.org/10.1080/14649357.2012.677124
Delgado-Ramos GC (2020) El peso de las ciudades mexicanas en un contexto de cambio climático: consumo de energía y materiales del Sistema Urbano Nacional. Pluralidad y Consenso 10(46):44–55
Delgado-Ramos GC (2021) Climate-environmental governance in the Mexico Valley Metropolitan Area: assessing local institutional capacities in the face of current and future urban metabolic dynamics. World 2(1):32–48. https://doi.org/10.3390/world2010003
Delgado-Ramos GC, Guibrunet L (2017) Assessing the ecological dimension of urban resilience and sustainability. Int J Urban Sustain Dev 9(2):151–169. https://doi.org/10.1080/19463138.2017.1341890
Delgado-Ramos GC, López García D (eds) (2020) Las ciudades ante el COVID-19 : nuevas direcciones para la investigación urbana y las políticas públicas. Plataforma de Conocimiento para la Transformación Urbana, Mexico. https://doi.org/10.5281/zenodo.3894075
Delgado-Ramos GC, Aragón-Durand F, Di Bella J, Franco C, Henríquez Ruiz C, Hernández Pedraza GC (2020) Urban and rural settlements. In: Moreno JM, Laguna-Defior C, Barros V, Bilbao B, Calvo Buendía E, Marengo JA, Oswald Spring Ú (eds) Adaptation to climate change risks in Ibero-American countries — RIOCCADAPT report. McGill, Madrid, pp 497–540
Eakin H, Bojórquez-Tapia LA, Janssen MA, Georgescu M, Manuel-Navarrete D, Vivoni ER et al (2017) Urban resilience efforts must consider social and political forces. Proc Natl Acad Sci U S A 114(2):186–189. https://doi.org/10.1073/pnas.1620081114

Egerer M, Haase D, Mcphearson T, Frantzeskaki N, Andersson E, Nagendra H, Ossola A (2021) Urban change as an untapped opportunity for climate adaptation. Urban Sustain 1(22):1–9. https://doi.org/10.1038/s42949-021-00024-y
Eisenmenger N, Ramos Martín J, Schandl H (2007) Análisis del metabolismo energético y de materiales de Brasil, Chile y Venezuela. Revista Iberoamericana de Economía Ecológica 6:17–39
ERN (n.d.) Estrategia para el aseguramiento colectivo de los inmuebles privados de Manizales. Evaluación de Riesgos Naturales. Consultores en Riesgos y Desastres. Online: http://idea.manizales.unal.edu.co/gestion_riesgos/descargas/transferencia/04info~1.pdf
Fahlberg A, Vicino TJ, Fernandes R, Potiguara V (2020) Confronting chronic shocks: social resilience in Rio de Janeiro's poor neighborhoods. Cities 99:102623. https://doi.org/10.1016/J.CITIES.2020.102623
Few R, Ramírez V, Armijos MT, Zambrano Hernández LA, March H (2021) Moving with risk: forced displacement and vulnerability to hazards in Colombia. World Dev 144:105482
Filgueira F, Galindo LM, Giambruno C, Blofield M (2020) América Latina ante la crisis del COVID-19 Vulnerabilidad socioeconómica y respuesta social. Comisión Económica para América Latina y el Caribe (CEPAL), Santiago
Filion P (2013) Fading resilience? Creative destruction, neoliberalism and mounting risks. Sapiens [Online] 6(1)
Fleischhauer M (2008) The role of spatial planning in strengthening urban resilience. Springer, Dordrecht, pp 273–298. https://doi.org/10.1007/978-1-4020-8489-8_14
Foster JB (2011) The ecological rift. Capitalism's war on the earth. Monthly Review Press, USA
Foster N (2020) Rethinking the right to the city: DIY urbanism and postcapitalist possibilities. rethinking marxism. J Econ Cult Soc 32(3):310–329
García Acosta V, Audefroy JF, Briones F (2012) Estrategias sociales de prevención y adaptación. CONACYT_CIESAS, Mexico City. Online: https://www.desenredando.org/public/2013/2013_estrategias_sociales_CIESAS.pdf
Gazzotti P, Emmerling J, Marangoni G, Castelletti A, van der Wijst K, Hof A, Tavoni M (2021) Persistent inequality in economically optimal climate policies. Nat Commun 12:3421
Glitz Mayrink P, López Javier SE, Quimbayo-Ruiz GA, Vásquez F (2021) Miradas a la ecología política de la urbanización desde Suramérica. In: Ariza-Montobbio P, Carrión A (eds) Ecología política urbana ante el cambio climático. FLACSO Ecuador- IDRC, p 274
Godschalk DR (2003) Urban hazard mitigation: creating resilient cities. Nat Hazards Rev 4(3):136–143. https://doi.org/10.1061/(asce)1527-6988(2003)4:3(136)
Guivarch C, Taconet N (2020) Global inequalities and climate change. Revue de L'OFCE 165(1):35–70
Gunderson LH, Holling CS (2002) Panarchy : understanding transformations in human and natural systems. Island Press, Washington, DC. http://www.loc.gov/catdir/enhancements/fy0666/2001006112-d.html
Guneralp B, Delgado-Ramos GC (2018) Food-energy-water security in a transborder urban context. Texas A&M-CEIICH-CONACYT. Research proposal 14210 of the award N° 2018-014.1
Harvey D (2008) The right to the city. New Left Rev 53. Online: https://newleftreview.org/issues/ii53/articles/david-harvey-the-right-to-the-city
Harvey D (2012) Rebel cities. From the right to the city to the urban revolution. Verso, London/New York
Henríquez C, Romero H (2019) Urban climates in Latin America. Springer International Publishing. https://doi.org/10.1007/978-3-319-97013-4
Hepburn C, O'Callaghan B, Stern N, Stiglitz J, Zenghelis D (2020) Will COVID-19 fiscal recovery packages accelerate or retard progress on climate change? Oxf Rev Econ Policy 36(Supplement_1):S359–S381. https://doi.org/10.1093/oxrep/graa015
Hornborg A (1998) Towards an ecological theory of unequal exchange: articulating world system theory and ecological economics. Ecol Econ 25(1):127–136. http://www.sciencedirect.com/science/article/pii/S0921800997001006

Hornborg A (2009) Zero-sum world: challenges in conceptualizing environmental load displacement and ecologically unequal exchange in the world-system. Int J Comp Sociol 50(3–4):237–262. https://doi.org/10.1177/0020715209105141
Houston D, Hillier J, MacCallum D, Steele W, Byrne J (2018) Make kin, not cities! Multispecies entanglements and 'becoming-world' in planning theory. Plan Theory 17(2):190–212. https://doi.org/10.1177/1473095216688042
ICLEI (2016) Guía de Acción Local por el Clima. ICLEI (International Council for Local Environmental Initiatives)-Gobiernos Locales Por la Sostenibilidad, São Paulo
IPCC (2014) Cambio climático 2014: Impactos, adaptación y vulnerabilidad – Resumen para responsables de políticas. In: Field CB, Barros VR, Dokken DJ, Mach KJ, Mastrandrea MD, Bilir TE et al (eds) Contribución del Grupo de trabajo II al Quinto Informe de Evaluación del Grupo Intergubernamental de Expertos sobre el Cambio Climático. Organización Meteorológica Mundial, Ginebra
IPCC (2018) Annex I: glossary [Matthews JBR (ed)]. In: Masson-Delmotte V, Zhai P, Pörtner H-O, Roberts D, Skea J, Pirani M IGA, Moufouma-Okia W, Péan C, Pidcock R, Connors S, Matthews JBR, Chen Y, Zhou X, Gomis MI, Lonnoy E, Maycock T, Tignor M, Waterfield T (eds) Global warming of 1.5°C. In Press. Available online: https://www.ipcc.ch/sr15/chapter/glossary/
IPCC (2021) Summary for Policymakers. In: Climate Change 2021: The Physical Science Basis. Contribution of Working Group I to the Sixth Assessment Report of the Intergovernmental Panel on Climate Change [Masson-Delmotte, V., P. Zhai, A. Pirani, S.L. Connors, C. Péan, S. Berger, N. Caud, Y. Chen, L. Goldfarb, M.I. Gomis, M. Huang, K. Leitzell, E. Lonnoy, J.B.R. Matthews, T.K. Maycock, T. Waterfield, O. Yelekçi, R. Yu, and B. Zhou (eds.)]. Cambridge University Press, Cambridge, United Kingdom and New York, NY, USA, pp. 3–32, https://doi.org/10.1017/9781009157896.001
IRP (2020) Resource efficiency and climate change: material efficiency strategies for a low-carbon future. International Resource Panel- UNEP, Nairobi
Kaika M (2017) 'Don't call me resilient again!' : the New Urban Agenda as immunology … or … what happens when communities refuse to be vaccinated with 'smart cities' and indicators. Environ Urban 29(1):89–102. https://doi.org/10.1177/0956247816684763
Lampis A (2013) Editorial: ciudad y resiliencia: una cuestión al estilo de Escher. Territorios 28:9–19. http://www.redalyc.org/articulo.oa?id=35728173001
LAWG (2020) A roadmap for transforming relations between the United States & the Northern Countries of Central America. Latin AmericaWorking Group Education Fund. Online: https://www.lawg.org/wp-content/uploads/LAWG-Central-America-recommendationsFIN2.pdf
Lefebvre H (1991) The production space. Blackwell Publishing, Maiden
Little A, Lloyd M (eds) (2009) The politics of radical democracy. Edimburgh University, Edinburgh
MacKinnon D, Derickson KD (2013) From resilience to resourcefulness. Prog Hum Geogr 37(2):253–270. https://doi.org/10.1177/0309132512454775
Marcuse H (1971) One-dimensional man: studies in the ideology of advanced industrial society, 2nd edn. Routledge, London
Meerow S, Newell JP, Stults M (2016) Defining urban resilience: a review. Landsc Urban Plan. Elsevier B.V. 147:38–49. https://doi.org/10.1016/j.landurbplan.2015.11.011
Meerow S, Pajouhesh P, Miller TR (2019) Social equity in urban resilience planning. Local Environ 24(9):793–808. https://doi.org/10.1080/13549839.2019.1645103
Mehmood A (2015) Of resilient places: planning for urban resilience. Eur Plan Stud 24(2):407–419. https://doi.org/10.1080/09654313.2015.1082980
Mendes FH, Romero H, Ferreira da Silva Filho D (2020) Cambio Climático adverso provocado por la urbanización sin planificación ni evaluación ambiental en Santiago de Chile. Rev Geogr Norte Gd 77:191–210. https://doi.org/10.4067/s0718-34022020000300191
Méndez R (2012) Ciudades y metáforas: sobre el concepto de resiliencia urbana. Ciudades y Territorios. Estudios Territoriales XLIV(172): 215–231

Meriläinen E (2020) The dual discourse of urban resilience: robust city and self-organised neighbourhoods. Disasters 44(1):125–151. https://doi.org/10.1111/DISA.12367
Metzger P, Robert J (2013) Elementos de reflexión sobre la resiliencia urbana: usos criticables y aportes potenciales. Territorios 28:21–40
Mitchell F, Ayala E, Della-Ceca L, Gimeno M, Aguilar JM (2019) Planes Locales de Acción Climática: Aportes para su elaboración y lecciones aprendidas de la experiencia de la Red Argentina de Municipios frente al Cambio Climático. Buenos Aires, Argentina
Moreno JM, Laguna Defior C, Barros V, Calvo Buendía E, Marengo JA, Spring O (2020) Adatpation to climate chane risks in Ibero-American countries. RIOCCADAPT report. McGraw Hill, Madrid
OECD (2020) Development Co-operation report 2020: learning from crises, building resilience. OECD Publishing, Paris. Online: https://www.oecd-ilibrary.org/docserver/f6d42aa5-en.pdf?expires=1630954609&id=id&accname=guest&checksum=644CCB36A930CAD939CBFFD0554F209C
Oke C, Bekessy SA, Frantzeskaki N, Bush J, Fitzsimons JA, Garrard GE et al (2021) Cities should respond to the biodiversity extinction crisis. npj. Urban Sustain 1(1):1–4. https://doi.org/10.1038/s42949-020-00010-w
Pelling M, Chow WTL, Chu E, Dawson R, Dodman D, Fraser A et al (2021) A climate resilience research renewal agenda: learning lessons from the COVID-19 pandemic for urban climate resilience. Clim Dev:1–8. https://doi.org/10.1080/17565529.2021.1956411
Rebotier J, López Peláez J, Pigeon P (2013) Las paradojas de la resiliencia: miradas cruzadas entre Colombia y Francia. Territorios 28(28):127–145. https://www.redalyc.org/articulo.oa?id=35728173006
Rigolon A, Browning M, Lee K, Shin S (2018) Access to urban green space in cities of the Global South: a systematic literature review. Urban Sci 2(3):67. https://doi.org/10.3390/urbansci2030067
Rivero-Villar A, Vieyra Medrano A (2021) Governance for urban resilience in popular settlements in developing countries: a case-study review. Clim Dev 14(3):208–221. https://doi.org/10.1080/17565529.2021.1906203
Robertson SA (2018) Rethinking relational ideas of place in more-than-human cities. Geogr Compass 12(4):e12367. https://doi.org/10.1111/gec3.12367
Romero H, Fuentes C, Smith P (2010) Ecología política de los riesgos naturales y de la contaminación ambiental en Santiago de Chile: necesidad de justicia ambiental. Scripta Nova. Revista Electrónica de Geografía y Ciencias Sociales XIV(331(52))
Romero-Lankao P, Gnatz D, Wilhelmi O, Hayden M (2016) Urban sustainability and resilience: from theory to practice. Sustainability 8(12):1224. https://doi.org/10.3390/su8121224
Rossi U, Vanolo A (2015) Urban neoliberalism. In: International encyclopedia of the social & behavioral sciences, vol 24, 2nd edn, pp 846–853
Samaniego P, Vallejo MC, Martínez Alier J (2017) Commercial and biophysical deficits in South America, 1990–2013. Ecol Econ 133:62–73
Sandoval V, Sarmiento JP (2020) A neglected issue: informal settlements, urban development, and disaster risk reduction in Latin America and the Caribbean. Disaster Prev Manag Int J 29(5):731–745. https://doi.org/10.1108/DPM-04-2020-0115
Selee A, Tanco A, Ruiz Soto A (2021) Setting the course toward sustainable migration cooperation: levering bilateral opportunities. Wilson Center. Online: https://www.wilsoncenter.org/sites/default/files/media/uploads/documents/Setting%20the%20Course%20Toward%20Sustainable%20Migration%20Cooperation_Levering%20Bilateral%20Opportunities.pdf
Shingne MC (2020) The more-than-human right to the city: a multispecies reevaluation. J Urban Aff:1–19. https://doi.org/10.1080/07352166.2020.1734014
Smith N (2010) Uneven development: nature, capital and the production of space. London/New York, Verso

Solecki W, Delgado Ramos GC, Roberts D, Rosenzweig C, Walsh B (2021) Accelerating climate research and action in cities through advanced science-policy-practice partnerships. NPJ Urban Sustain 1(1):1–8. https://doi.org/10.1038/s42949-021-00015-z

Svampa M (2019) Neo-extractivism in Latin America. Cambridge elements. Cambridge University Press, Cambridge

Swyngedouw E, Ernstson H (2019) O tempora! O mores! interrupting the anthropo-obscene. In: Urban political ecology in the anthropo-obscene interruptions and possibilities. Routledge, Abingdon/New York, pp 25–47

UNEP (2021) El Peso de las Ciudades en América Latina y el Caribe: Requerimientos Futuros de Recursos y Potenciales Rutas de Actuación. Delgado Ramos GC, Martino D (eds). UNEP. Nairobi

UNEP, UN-HABITAT (2021) Global environment for cities: towards green and just cities. Nairobi

UN-HABITAT (2011) Global report on human settlements 2011. Cities and climate change. Earthscan, London, Washington. Online: https://unhabitat.org/sites/default/files/download-manager-files/Cities%20and%20Climate%20Change%20Global%20Report%20on%20Human%20Settlements%202011.pdf

UN-HABITAT (2015) HABITAT III issue papers 22 – informal settlements. New York

UN-HABITAT (2016) World cities report 2016: urbanization and development – emerging futures. New York

UN-HABITAT (2021) Cities and pandemics: towards a more just, green and healthy future. UNHABITAT, Nairobi. Online: https://unhabitat.org/sites/default/files/2021/03/cities_and_pandemics-towards_a_more_just_green_and_healthy_future_un-habitat_2021.pdf

Staden R van (2014) Climate change: implications for cities. key findings from the Intergovernmental Panel on Climate Change Fifth Assessment Report. University of Cambridge. ICLEI Local Governments for Sustainability

WCRP (2019) Global research and action agenda on cities and climate change science cities and climate change, vol 13. World Climate Research Programme

Wolff E (2020) Uncharted terrain: the clash between “risk management” and informal urbanisation. Urban Transcrip J 3(3)

Yang Q, Yang D, Li P, Liang S, Zhang Z (2021) Resilient city: a bibliometric analysis and visualization. Discret Dyn Nat Soc 2021. https://doi.org/10.1155/2021/5558497

Zeiderman A (2016) Adaptative publics: building constituencies in Bogotá. Publ Cult 28(2(79)):389–413

Part III
Bottom-Up and Co-Produced Resilience

Chapter 6
Nature-Based Climate Solutions in European Schools: A Pioneering Co-designed Strategy Towards Urban Resilience

Francesc Baró, David A. Camacho, Carmen Perez del Pulgar, Isabel Ruiz-Mallén, and Pablo García-Serrano

Abstract As cities around the world are increasingly facing the impacts of climate change, a growing number of municipalities are leading collaborative projects to adapt schools to rising temperatures. The implementation of nature-based solutions (NBS) is highlighted as an important component to be included in these initiatives given their multifunctional and cost-effective character. However, the challenges and upscaling opportunities of these pioneering projects are still not well understood nor systematically studied on a comparative basis. This study explores and compares three European pilot nature-based projects aiming to make schools more

---

F. Baró (✉)
Department of Geography and Department of Sociology, Vrije Universiteit Brussel (VUB), Brussels, Belgium

Institute of Environmental Science and Technology (ICTA), Universitat Autònoma de Barcelona, Bellaterra, Barcelona, Spain
e-mail: francesc.baro@vub.be

D. A. Camacho
Institute of Environmental Science and Technology (ICTA), Universitat Autònoma de Barcelona, Bellaterra, Barcelona, Spain

C. Perez del Pulgar
Environmental Policy Department, Helmholtz Centre for Environmental Research – UFZ, Leipzig, Germany

Department of Political Science, Friedrich-Schiller-University Jena, Jena, Germany

Institute of Environmental Science and Technology (ICTA), Universitat Autònoma de Barcelona, Bellaterra, Barcelona, Spain

I. Ruiz-Mallén
Estudis de Psicologia i Ciències de l'Educació and Internet Interdisciplinary Institute (IN3), Universitat Oberta de Catalunya (UOC), Barcelona, Spain

P. García-Serrano
Serendipia Educativa, Madrid, Spain

I. Ruiz-Mallén et al. (eds.), *Urban Resilience to the Climate Emergency*,
The Urban Book Series, https://doi.org/10.1007/978-3-031-07301-4_6

resilient to climate change impacts while creating other co-benefits for children and the wider local community. These projects are *Oasis* (Paris), *Climate Shelters* (Barcelona) and *Care in School Environments* (Madrid). Building on a framework for assessing the co-benefits of urban NBS, the comparative analysis explores the selection criteria of schools and their equity implications, the multifunctional role of NBS beyond climate adaptation, the main aspects and challenges related to the co-design process and the subsequent project implementation and the potential for upscaling at the city level. Based on this exploration, we contend that nature-based climate adaptation projects in schools can be a spearhead for a wider community-based strategy towards urban resilience.

Keywords Nature-based solutions · Green resilient infrastructure · Children's geographies · Urban equity · Green schoolyards

6.1 Introduction

As cities around the world are increasingly facing the impacts of climate change, a growing number of municipal governments are fostering or directly leading the implementation of transformative co-produced projects to adapt schools to rising temperatures, including more intense urban heat islands. A recent report by the C40 Cool Cities Network (Vetter 2020) identified several reasons to make schools more resilient to climate change impacts, including: (1) children are a particularly vulnerable age group to extreme heat; (2) creating schoolyards with enough shaded areas encourages children to play outside and increases physical activity during recess time; (3) using schools as community hubs can bring climate awareness to families; and (4) transformed schools can act as *cool islands* or *climate shelters* that can be also used by neighbours after school hours or on weekends, especially during hot days. According to the report, an important component to be promoted in school climate adaptation projects is the implementation of *nature-based solutions* (NBS), such as tree plantations, green roofs and walls, water elements or educational and rain gardens. NBS are defined by the European Commission as *solutions that are inspired and supported by nature which are cost-effective, simultaneously provide environmental, social and economic benefits and help build resilience* (Wild et al. 2020 p. 3). Similarly, the International Union for the Conservation of Nature (IUCN) refers to NBS as *actions to protect, sustainably manage and restore natural and modified ecosystems in ways that address societal challenges effectively and adaptively, to provide both human well-being and biodiversity benefits* (Cohen-Shacham et al. 2019 p. 21). In general, the concept of NBS is considered as an umbrella term bringing together previously established ecosystem-based notions such as *ecosystem services*, *green and blue infrastructure* or *natural capital* (Nesshöver et al. 2017).

The implementation of NBS in school environments can thus play an important role towards rapid, systemic transitions leading to climate-resilient cities (de Coninck et al. 2018). Further, as reflected in the UN Sustainable Development Goal

(SDG) 11 (Target 11.7), cities should also ensure an equitable access to safe and inclusive green and public spaces, particularly in the case of children. However, previous research has shown a declining and unequal trend regarding children's use of and access to urban green spaces (Gidlow and Ellis 2011). Incorporating NBS in school settings seems a promising path to overcome disparities in residential access to urban nature and to boost multiple co-benefits ranging from climate change adaptation to enhanced health, wellbeing and learning (Chawla et al. 2014; van Dijk-Wesselius et al. 2018, 2020; Baró et al. 2021).

Most children spend a substantial share of their daily life in school settings. While generally most of the school time is spent indoors, children also use outdoor environments (typically school playgrounds) during recess or physical activity time, in some cases up to 10 h per week (Slater et al. 2012; Shape America 2016). Despite school grounds are still usually dominated by sport fields and paved surfaces, nature-based designs are increasingly seen and valued as innovative alternatives thanks to their multiple positive impacts on children. The transformation of school settings into *green environments* builds on a growing body of literature showing a strong positive association between (urban) nature and child overall wellbeing and health (Bell and Dyment 2008; Chawla 2015). Everyday access to school green and blue spaces has been associated with health benefits such as moderation of stress (Chawla et al. 2014; Akpinar 2016) and improvement of attention-deficit hyperactivity disorder's (ADHD) symptoms (Faber Taylor and Kuo 2011; Markevych et al. 2014). There is also evidence on the beneficial associations between higher exposure to urban vegetation and schoolchildren behavioural and cognitive development (Amoly et al. 2014; Dadvand et al. 2015; Pérez-del-Pulgar et al. 2021). Several studies have also assessed the impact of school green space on the academic performance of pupils, yet results are generally non-significant or mixed so far (Browning and Rigolon 2019; Kuo et al. 2021). Further, green schoolyards can overcome the classic design of playgrounds situating sport fields at the centre (generally used by boys) and smaller recreational areas in the surroundings (generally used by girls), which often results in a strong spatial gender hierarchy. In contrast, naturalized playgrounds can help diversifying play options and improve the social interaction between girls and boys (Equal Saree 2019) while establishing more supportive relationships and strengthening their emotional and relational wellbeing (Chawla et al. 2014). Green spaces in school environments are also increasingly seen and used as outdoor learning environments where children can develop environmental knowledge and (lifelong) affinity towards nature (Broom 2017; Rosa et al. 2018; van Dijk-Wesselius et al. 2020). Finally, tree canopy and other vegetated covers can have a substantial positive impact on urban schoolyards' microclimate, by enhancing thermal comfort and reducing heat stress perception of children (Antoniadis et al. 2020).

In Europe, recent nature-based climate initiatives to transform school settings are generally implemented through co-production approaches, that is, engaging factual and potential users of these spaces since the earliest stages of the design process. However, while the benefits of nature-based school transformations within the context of urban climate resilience are increasingly supported by scientific evidence as described above, the implementation challenges and upscaling opportunities of this

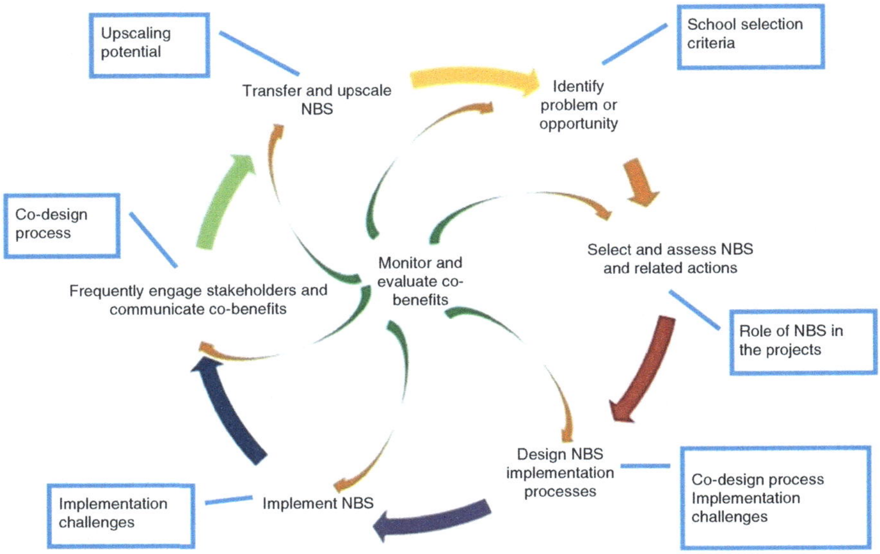

Fig. 6.1 Aspects assessed in the three nature-based climate resilience school projects. (Modified from Raymond et al. (2017), published under a Creative Commons license (CC BY 4.0))

kind of projects are still not well understood nor systematically studied on a comparative basis. In response to this knowledge gap, the goal of this chapter is to explore and compare three European pioneering pilot projects aiming to make schools more resilient to climate change impacts while creating healthier, more playful and pedagogical environments for children, mostly through nature-based interventions in their schoolyards. These projects are *Oasis* (implemented in Paris, France), *Climate Shelters* (Barcelona, Spain) and *Care in School Environments* (Madrid, Spain). Indeed, these cities represent large-size cities where the nature-based school interventions have been conducted in neighbourhoods with different socioeconomic and environmental conditions. Working in these settings allows for both unravelling the specificities of each particular context and finding common patterns.

The comparative analysis builds on the framework for assessing and implementing the co-benefits of NBS in urban areas developed by Raymond et al. (2017), focusing on the following aspects (see also Fig. 6.1): (1) selection criteria of schools and related equity implications; (2) the role of NBS in the projects; (3) participatory process and level of co-design achieved; (4) main barriers or challenges encountered during project implementation; and (5) potential for upscaling, understood as how successful initiatives testing new practices, services or governance approaches can be mainstreamed or brought to higher policy levels (Fastenrath et al. 2020).

6.2 Description of the Three Nature-Based Climate Resilience School Projects

6.2.1 Oasis (Paris)

The Oasis project (standing for Openness, Adaptation, Sensitization, Innovation and Social ties) involves the transformation of ten school playgrounds (including nursery, primary and secondary centres) across the city of Paris into *cool islands* using both grey (engineering) and green (nature-based) solutions. The project aims to address climate challenges such as heat stress mitigation (heatwaves and their exacerbation due to the urban heat island effect are identified as a major risk for the city) and stormwater control (including flooding risk) but also contribute to environmental awareness and social cohesion both across school communities and the wider neighbourhood community. The project was funded under the UIA programme (Urban Innovation Actions, part of the European Regional Development Fund – ERDF) with almost 5 M € and has been implemented during the period 2019–2021.

The initiative involves a cross-disciplinary collaboration between the municipality, local civic organizations, associations and research centres coordinated by the Resilience Office of the City of Paris (see also Fig. 6.2). Other City departments directly involved include the Department of Education and the Department of Architecture and their respective local divisions. The other partners include (1) the Council of Architecture, Urban Planning and the Environment, a local NGO responsible for the co-design and training process with the school communities; (2) the Paris Federation of the Education League, a platform of local associations that is responsible for the broader local community engagement process; (3) the French National meteorological and climatological centre, responsible for environmental monitoring and educational activities; (4) the Interdisciplinary Research Laboratory for Future Energies, responsible for the microclimatic and thermal evaluation; and (5) the Laboratory for Interdisciplinary Evaluation of Public Policies, at Sciences Po, responsible for the overall participatory, social impact and evaluation process. The funded project is part of a wider OASIS framework initiated as an action plan of the City's Resilient Strategy (Paris City Council 2018). The ultimate goal of this plan is to have all 760 Paris' public schoolyards transformed into *oases* by 2050.

6.2.2 Climate Shelters (Barcelona)

Climate Shelters (*Refugis Climàtics*) is a twin project of Oasis since it shares the same funding scheme (UIA initiative) and operational approach based on the transformation of 11 primary schools in Barcelona through climate-resilient green, blue and grey interventions. This pilot project involves creating more shaded spaces through greenery, incorporating water points, improving buildings' ventilation and

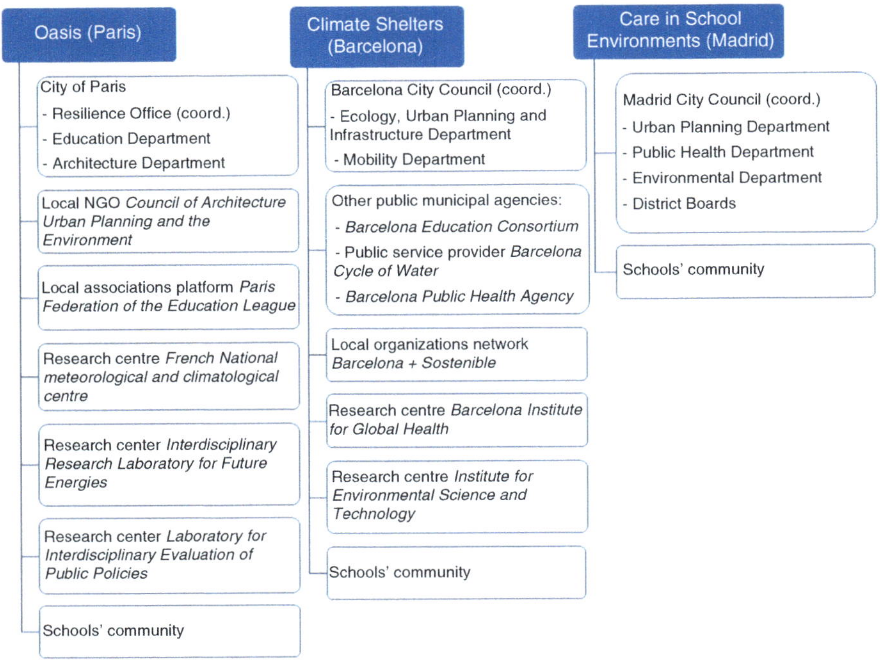

Fig. 6.2 Main stakeholders involved in the co-design process of nature-based climate resilience solutions at schools in each case study project. School community actors include school board, teachers, schoolchildren and parents

implementing permeable materials, among other interventions. The ultimate goal is that schools (both buildings and grounds) serve as cooling spaces to pupils, school staff and external visitors (through the programme *Patis escolars oberts al barri* – Schoolyards open to the neighbourhood), especially during hot days. It is also planned that participating schools will expand the environmental education approach in their curriculum thanks to the new nature-based interventions. The project started in 2019, and it is expected to finish in 2022, counting with a total budget of almost 5 M € (4 M € from the UIA programme and 1 M € provided by the Barcelona City Council).

Besides the municipality, the initiative involves other public agencies and research centres (see Fig. 6.2). It is led by the Ecology, Urban Planning, Infrastructures and Mobility Department of the Barcelona City Council, and the Barcelona Education Consortium collaborates with the management and implementation of the measures. The project is also embedded in the programme *Escoles + Sostenibles* (More Sustainable Schools), the educational branch of a wider network of public, private and civic organizations fostering environmental, social and economic sustainability in Barcelona (*Barcelona + Sostenible*). Other partners include the Barcelona Cycle of Water (public service provider) which provides expertise on water usage; the Public Health Agency of Barcelona and the

Barcelona Institute for Global Health, both evaluating the health impacts of the interventions; and the Institute for Environmental Science and Technology, assessing environmental quality aspects such as air pollution levels. The project is part of a wider line of action to prevent excessive heat through the identification and creation of climate shelter spaces included in the Barcelona Climate Plan 2018–2030 (Barcelona City Council 2018) and the Climate Emergency Declaration (2020). By the end of the project, it is expected to transform 3000 m^2 of schoolyards and to have clear guidelines for replicating the interventions in other schools in the city.

6.2.3 Care in School Environments (Madrid)

Care in School Environments (*Cuidados en Entornos Escolares*, own translation) was a project aiming to create a new school ground model by transforming the schoolyards of nursery and primary public schools in the municipality of Madrid. The project responded to the need to undertake a profound and holistic revision of the – so far dismissed – relevance of schoolyards for children's education, health and development and as spaces of opportunity for climate change adaptation interventions.

Since its inception in 2017, the project was characterized by having evolved without a specifically dedicated municipal budget. This fact led to a strong dependence on the voluntaristic motivation of the municipal officers of the Urban Planning, Public Health and Environmental departments and district boards with a direct influence on the development of the project (see Fig. 6.2). The actual project became possible thanks to the cross-departmental and district pooling of existing budgets and resources previously allocated to other goals (e.g. district level funds for school maintenance, climate change programmes, urban plans etc.). This budget pooling managed to allocate a total of 1 M € to schoolyard transformation without creating a dedicated new municipal budget line.

The project involved the transformation of three schoolyards and was implemented between 2017 and 2019. The selected schools were located in areas of priority intervention in regard to the major working lines of the three involved departments: (1) social vulnerability, identified in the Madrid Urban Regeneration Strategy of the Urban Planning Department (Madrid City Council 2016–2018); (2) climate vulnerability to extreme events, identified in the Air Quality and Climate Change Plan of the Environmental Department (Madrid City Council 2017); and (3) health vulnerability, identified in the broader City Public Health Program *Madrid a City that Cares* (Madrid City Council, 2017–2021). In addition, the selected schools were located in districts whose local board agreed to reorient the budget of the framework contracts for school maintenance towards school ground transformation works. The project also developed a preliminary City Schoolyard Intervention Strategy identifying the available environmental and play resources in and around all schools of Madrid (García-Serrano et al. 2017a).

6.3 Data Collection and Comparative Analysis

Information on the three case study projects was collected via semi-structured interviews with the main project coordinators and promoters and also complemented with institutional project documents available online.[1] Therefore, the comparative analysis is limited to the view of these institutional partners, and it does not reflect the perspectives or experiences of *end users* who participated in the co-design processes (e.g. school personnel, parents or children). For Oasis, interviewees included the Deputy Chief Resilience Officer of the City of Paris and an UIA Expert. In the case of Care in School Environments, we interviewed the Co-manager of the project from the General Directorate of Strategic Planning of the City of Madrid. Finally, the International Relations Project Coordinator and Head Architect Manager for the Ecology, Urban Planning and Mobility Area of the Barcelona City Council and the responsible for the *Escoles + Sostenibles* programme were interviewed as coordinators of Climate Shelters. All interviews were conducted between January and June 2021 via video-conference due to the Covid-19 pandemic restrictions. The guiding interview questions built on the five main analysis points mentioned in the introduction section as follows:

1. School selection criteria: What were the main selection criteria of schools (i.e. exposure to environmental harms such as pollution, heatwaves, equality/equity/justice criteria etc.)?
2. The role of NBS: What are the main NBS interventions included in the project (e.g. rain gardens, trees, blue spaces etc.)? For which challenges/goals these NBS elements have been mostly designed (e.g. heatwaves, runoff control, environmental education etc.)?
3. Participatory process: Who has been mainly involved in the participatory process of the project (e.g. school staff, parents etc.)? What was the role of schoolchildren during the process? What is the level of co-production/co-implementation achieved?
4. Implementation barriers/challenges: Which are the main barriers/obstacles faced during the implementation of the project (e.g. restricted timeframe, budget cuts, coordination between partners, Covid-19 crisis etc.)? How have been these barriers addressed (or not)?
5. Upscaling potential: Has the project made any relevant changes after/during its implementation that were not anticipated? Is the project planned to be upscaled at city level? How?

The comparative analysis was based on a manual coding of the five analysis points as reflected in a synthesis of the interview transcripts and the reviewed documents.

[1] Oasis: www.uia-initiative.eu/en/uia-cities/paris-call3; Climate Shelters: https://www.uia-initiative.eu/en/uia-cities/barcelona-call3; Care in School Environments: https://madridsalud.es/cuidado-de-los-espacios-publicos-de-los-colegios/

6.4 Results

6.4.1 Selection Criteria of Schools and Equity Implications

Several school selection criteria were considered across the three projects (see an overview in Table 6.1). The shared funding scheme of Oasis in Paris and Climate Shelters in Barcelona is probably the reason that both projects applied similar criteria. For instance, the two cases had a similar budget to transform 10–11 schools, and both decided to apply a balanced geographical distribution across the different city districts (at least one selected school per district). In contrast, Care in School Environments, in Madrid, with a more limited funding scheme and scope, was only implemented in three schools located in priority areas of intervention for its social, environmental and health vulnerabilities with already approved operational budgets.

Criteria related to school type (public vs private or charter), size and accessibility were key in the selection process. In all cases, eligible schools had to be publicly funded and giving nursery and/or primary education (except for Oasis where two secondary schools – *collèges* – were also included). In Barcelona, schools with a higher number of pupils were also prioritized. In Paris, it was mandatory that selected schools had a schoolyard with direct street access to make it accessible to external users. Similarly, in Barcelona, it was required that schools were part of the *Schoolyards open to the neighbourhood* programme, which promotes the public use of schoolyards after school hours by neighbourhood residents. Further, the selection of schools was also based on the support/interest of district authorities (Oasis and

Table 6.1 Summary of the school selection criteria in each case study project. Note: two checks indicate that the criterium was mandatory or very relevant, and one check indicates that it was supplementary or secondary

Criteria	Oasis (Paris)	Climate Shelters (Barcelona)	Care in School Environments (Madrid)
Equal geographical distribution (across city districts)	✓✓	✓✓	
Pre-existing intervention plans (with secured budget)			✓✓
District authority and/or school support	✓	✓✓	✓
Publicly funded school	✓✓	✓✓	✓✓
Schoolyard with direct access to street (to make it accessible to external visitors)	✓✓	✓✓	✓
Poor condition of the schools (building rehabilitation needed; low greenery)	✓	✓	✓
High exposure to environmental/climate harms (e.g. urban heat island, air pollution etc.)	✓	✓	✓
Number of pupils and other potential beneficiaries		✓	

Care in School Environments) and/or the schools themselves (Spanish cases). In fact, the schools selected in the Climate Shelters project had to follow an open application process where all selection criteria were evaluated based on a comprehensive threshold/points system.[2]

Finally, poor condition of schools in terms of the buildings' state (e.g. low thermal isolation) or lack of greenery in the schoolyards, together with a high exposure to environmental harms such as urban heat island effects or air pollution, were considered also as selection criteria in all cases, but with a more secondary weight.

6.4.2 The Role of Nature-Based Solutions in the Projects

The role of NBS is prominent in all the projects thanks to its multifunctionality in terms of climate adaptation (heat mitigation and stormwater runoff control), biodiversity, environmental education and health benefits. The main NBS intervention implemented in the three cases, with a clear focus on climate adaptation, was the replacement of impervious schoolyard pavements by natural and permeable materials (e.g. wood chip, sand areas, natural soil) and the construction of diverse topologies (e.g. small hills) and play structures. All projects included different types of green infrastructure (see also Fig. 6.3) such as gardens with native species to enhance biodiversity (e.g. Mediterranean gardens in Barcelona and Madrid), green walls, vegetated pergolas and increased tree canopy cover to create shaded spaces (e.g. a total of 74 trees have been planted in Climate Shelters). The creation of rain gardens and other sustainable drainage systems (including stormwater storage tanks) was especially common in Oasis and Care in School Environments. Blue infrastructure was also relevant in the three projects. Traditional and multifunctional (e.g. sprayer showers) fountains were implemented in all cases. Other innovative and singular water elements directly addressing environmental education goals were also built, such as the case of pedagogical rivers and small ponds in Paris, water-based game spaces in Barcelona and drinking fountains in Madrid (see a summary in Table 6.2).

Besides green and blue infrastructure, projects also included *grey* interventions, mostly oriented to increase the energy efficiency of school buildings (including the installation of solar panels), to enhance indoor air circulation and to create shaded areas.

[2] See: https://www.uia-initiative.eu/en/news/adapting-barcelona-climate-change-multicriteria-approach-selection-schools-be-converted

Fig. 6.3 Pictures of nature-based solutions implemented in the three case study cities. Upper photo: schoolyard of an Oasis project school (Paris) with permeable pavement, trees and pedagogical stream (Source: Laurent Bourgogne, City of Paris). Middle photo: schoolyard of Can Fabra

Table 6.2 Summary of the main green and blue infrastructure elements included in the case study projects

Main green and blue infrastructure elements	Oasis (Paris)	Climate Shelters (Barcelona)	Care in School Environments (Madrid)
Natural and permeable pavements (e.g. wood chip, sand areas) with diverse topologies and play structures	✓	✓	✓
Gardens with native species	✓	✓	✓
Green walls and vegetated pergolas	✓	✓	✓
Increased tree canopy cover (shade areas)	✓	✓	✓
Rain gardens and other sustainable drainage systems	✓		✓
Traditional and multifunctional fountains	✓	✓	✓
Pedagogical river and other singular water elements	✓	✓	

6.4.3 Co-design Process

All projects invested a substantial share of their resources in implementing a comprehensive participatory process or co-design strategy mostly focused on the transformation of the schoolyards (as the interventions in the buildings were more technical). Table 6.3 provides an overview of the main aspects related to these processes. In all cases, these strategies primarily aimed to engage with different stakeholders/groups of the school community, including schoolchildren, school staff (teachers, directors etc.), extracurricular instructors and also parents.

The engagement processes were mostly implemented through a series of workshops facilitated by experienced knowledge brokers. In the case of the project Care in School Environments, the process also included a preliminary phase of Participatory Action Research (PAR, see Pain et al. 2012) – with a core group of participants representing the educational community – and the use of surveys. The workshops were designed based on the target group (pupils, staff or parents) in Oasis, whereas Climate Shelters and Care in School Environments also organized sessions with mixed groups (i.e. all actors together).

Fig. 6.3 (continued) school (Climate Shelters, Barcelona) with wood structures, pergolas and vegetated walls (*Pati de l'Escola Can Fabra* by Òscar Giralt and Gerència d'Àrea d'Ecologia Urbana licensed under CC-BY-NC-ND). Lower photo: schoolyard of Ramón María del Valle Inclán School (Care in School Environments, Madrid) with impervious pavement removed, natural soil, new permeable pavements, rainwater harvesting and planted trees and bushes. (Source: Google Earth 6.0, (2008) 3D Buildings data layer. Available at: http://www.google.com/earth/index.html [Accessed 20/06/2021] & Ramon Maria del Valle Inclán School Schoolyard Core Group). All images are published with the permission of the owner(s)

Table 6.3 Summary of the main aspects related to the co-design process in the case study projects

Main aspects of the co-design process	Oasis (Paris)	Climate Shelters (Barcelona)	Care in School Environments (Madrid)
Participants in the main co-design process	School community (pupils, staff, parents etc.)	School community + wider local community	School community + wider local community + technical council staff
Tools used during the co-design process	Workshops	Workshops	PAR methodology, survey, workshops
Main stages of the co-design process	Diagnosis, co-design	Diagnosis, co-design, evaluation	Diagnosis, co-design, evaluation
Additional sustainability/resilience educational events	Specific raising-awareness workshops	City sustainability programme in schools	City sustainability programme in schools
Participation of external stakeholders (beyond school community)	Yes, *citizen assemblies*	Yes, as part of main co-design process	Yes, as part of main co-design process

All projects followed two main steps during the co-design process: (1) diagnosis (current state, uses of the schoolyard and identification of needs or deficiencies) and (2) co-design (discussing, sharing and co-designing ideas for the transformation of the schoolyard). Building on the outputs of the co-design stage, the technical departments of each City Council developed the final executive plans for each school. It is worthy to mention that in the case of Climate Shelters and Care in School Environments, the plan was not considered definitive until the school had approved it. In Barcelona and Madrid, a last follow-up step was included in the form of evaluation sessions (and a report in the case of Madrid) intended to estimate the level of satisfaction of the school community with the implemented interventions. In addition, a second impact evaluation is being carried out in Madrid as of 2021 (3 years after the interventions) using different qualitative methods (focus groups, interviews, questionnaires etc.). The Oasis project also organized *raising-awareness* workshops for schoolchildren with educational materials about climate change to motivate and boost their engagement in the co-design process. In Madrid and Barcelona, students from some of the transformed schools (all in the case of Barcelona) were already aware and engaged in climate change activities as a result of their previous participation in educational programmes for environmental sustainability coordinated by the City Councils (*Educar hoy por un Madrid más sostenible* and *Escoles + Sostenible* respectively).

Besides the school community, the Climate Shelters and Care in School Environments projects also included the participation of other community stakeholders (e.g. neighbourhood associations, sport organizations that make use of school facilities after school hours etc.) as part of the main co-design process (including the preliminary PAR phase in the case of Madrid). In contrast, the Oasis project included a distinct participatory process oriented to the broader local community (i.e. beyond the school actors). The goal behind this process was to engage with potential users of the schoolyards after school hours since the aim of the project also was to open them to the neighbourhood as additional public spaces. A series of *citizen assemblies* were organized (mostly online due to Covid-19 pandemic) in order to identify concerns and opportunities of the schoolyards as shared community spaces after school hours (e.g. potential activities allowed). However, this process was not coordinated with the co-design school community phase, so the final schoolyard plans did not reflect the outputs at this broader level.

6.4.4 Main Challenges Encountered During Project Implementation

As usual, all projects faced several barriers during their implementation (see a summary in Table 6.4), but one overarching barrier was directly related to the project capacity for entailing transformative processes. All project promoters stated explicitly or implicitly the *reluctance to change* expressed by some actors. For example, the implementation of NBS in the schoolyards was an unusual intervention to be executed by project engineers and perceived as *dirty* or *unsafe* by some adults (parents and school staff) in the Oasis project. In general, the participatory processes and the engagement with the school community (especially teachers) played a key role in order to overcome this reluctance. In Oasis, several *peer learning* sessions were also organized consisting of visiting schools (from other countries) with similar projects already implemented. Both school personnel (e.g. directors, teachers) and technical staff (architects, engineers) could talk to their peers involved in other school transformation projects and share experiences. In Madrid, the inclusive and

Table 6.4 Summary of the main challenges encountered (first column) and strategies followed to address or mitigate them (other columns)

	Strategies to address or mitigate the challenges		
Main challenges	Oasis (Paris)	Climate Shelters (Barcelona)	Care in School Environments (Madrid)
Reluctance to change by (some) stakeholders	Co-design process and *peer learning* sessions	Co-design process	Co-design process
Malpractice in project implementation	Not reported	Not reported	Close monitoring of construction works
Tight implementation timeframe, bureaucratic barriers, silo thinking	High political support	High political support	High degree of voluntaristic work Pooling of existing municipal budgets
Covid-19 pandemic	Online participation	Online participation and digital *tours*	NA (pre-pandemic implementation)

robust participatory process (characterized by the PAR methodology) proved to be decisive in minimizing conflict with the school community and avoiding the rejection of the interventions. However, due to the innovative nature of some interventions, the monitoring of the construction works had to be very thorough in order to avoid *malpractice* by the construction companies, i.e. deviation from what was decided in the co-design process and implementation of *standard* solutions.

In all cases, the alignment of the project timeframe with the schools' rhythms and availability was also challenging. The design of the solutions had to be done in a very short period of time, which didn't allow to tailor them as initially desired. Similarly, its implementation had to be done mostly during the summer and after school hours in order to avoid interference with the school activities. Moreover, in Barcelona, some school buildings were classified as historical heritage which required a longer bureaucratic process before the works could be started. These concerns were expressed as follows: *in terms of works, projects had to be drafted in a month and a half, public tenders had to be issued in a month and executed in barely 2 months... the results could have been better with more time. What we did in 8 or 9 months should have been done in a year and a half, so that the final solutions would have been more adequate, fairer* (Co-manager of Climate Shelters, Barcelona).

High-level political support and/or will was highlighted as a key enabler or barrier for the projects' implementation. In Oasis and Climate Shelters, a strong political support fostered the coordination between the different city departments involved in the project, avoiding silo thinking and pushing for the timely implementation of the interventions. On the other hand, Care in School Environments shows how a bottom-up project can have a greater transformative potential adapted to the communities' needs but face difficulties for upscaling due to the lack of political support (and budgetary limitations).

Finally, the outbreak of the Covid-19 pandemic impacted the projects in several ways. For instance, in Paris the restrictions associated with the pandemic delayed the transformation of one school and the opening of the schoolyards to the broader community. In Barcelona, the pandemic changed the uses and dynamics of the schoolyards. For instance, during winter, the classrooms were forced to keep the windows always open, making it impossible to measure the effectiveness of the new school ventilation systems. All this resulted in the delay or difficulty in evaluating the expected social, health and environmental changes as initially designed. Moreover, the dynamics of the workshops were affected by social distancing. For example, in the Climate Shelters project it was planned to organize schoolyard tours where the architects and project technicians could explain to the community the different infrastructures to be implemented. However, due to the pandemic, these activities had to be done using photographs and plans via video calls, making the communication process more difficult. However, the Covid-19 pandemic was also seen as an opportunity for upscaling this kind of initiatives, as it is explained in the next subsection. In Madrid, the project was implemented before the pandemic, but participant schools have recently reported to have been in a better position to adapt to Covid-related school protocols thanks to the school ground transformations (Co-manager of the Care in School Environments, Madrid).

6.4.5 Potential for Upscaling

Oasis and Climate Shelters are pilot projects which will be replicated in other schools according to municipal plans. Actually, upscaling is already ongoing in Paris (with municipal funds), and all public schools are expected to be part of the Oasis framework by 2050. The City has also been contacted by many French municipalities interested in replicating the project in their own schools. In Barcelona, the plan is to transform at least ten schoolyards per year considering the processes and solutions implemented in the Climate Shelters projects. This will be operationalized through a stable action programme called *Transformem els patis escolars* (Transforming schoolyards). In both cities, the projects are backed by municipal climate adaptation policies (Barcelona Climate Plan 2030 and Paris Resilience Strategy). In Madrid, there are no current plans to upscale the project due to lack of political support from the current municipal government (which changed from a left-wing coalition towards a right-wing conservative coalition shortly after the implementation of the project). Nevertheless, the pioneering municipal public schoolyard intervention strategy deployed by Care in School Environments remains as an important legacy for school ground transformation, as shown by the widely referenced School Environments Design Guide (García-Serrano et al. 2017b).

Despite the abovementioned negative impacts of the Covid-19 pandemic, the Oasis project coordinator considered that it will also enhance the acceptability of this kind of projects by the (school) communities: *these projects don't seem now like something weird, but as something needed. I anticipate that the transformation of schoolyards into resilient, green, sustainable places will evolve globally* (Oasis project manager, Paris). The results of the research evaluations being carried out as of 2021 (in terms of climate, health or social benefits) will be also crucial to support upscaling efforts in these cities and others. Another point expressed by the Care in School Environments representative is that in order to make these investments cost-effective (also considering the higher maintenance costs), the transformed schoolyards should be publicly accessible beyond school hours: *the large investment required to adapt schools to climate change and update their space to a contemporary pedagogical model makes sense if they represent a leisure infrastructure for the neighbourhood beyond school hours* (Co-manager of the Care in School Environments, Madrid). This is already done in Barcelona and planned in Paris.

6.5 Discussion and Conclusion

Growing evidence on the multiple benefits of nature contact for the overall development and wellbeing of schoolchildren have supported the implementation of green school(yard) transformations during the last decades, especially in countries like Canada, the USA or the UK (Bell and Dyment 2008). School greening initiatives are gaining new momentum as a way to make school settings more resilient to climate change impacts through the increasingly mainstreamed concepts of urban green infrastructure (Demuzere et al. 2014) and NBS (Kabisch et al. 2017). In this chapter, we have examined three recently implemented pilot projects in different European cities to better understand the challenges that these co-produced initiatives have faced but also the potential opportunities for upscaling.

The first aspect analysed was the school selection criteria. Due to limited budgets and complex (co-)funding strategies, the number of transformed schools was restricted to 3 centres in Madrid, 10 in Paris and 11 in Barcelona. Despite climate resilience was highlighted as the main challenge to be addressed by the projects (especially for Climate Shelters and Oasis), a higher exposure to climate harms (e.g. urban heat island) was not considered as a key criterion for school selection. Instead, aspects related to school type, condition, size, accessibility or institutional support (by the schools themselves or by municipal authorities) played a more relevant role. A balanced geographical distribution (at least one selected school per district) was also a mandatory prerequisite in Barcelona and Paris, regardless of the socioeconomic characteristics of each district. Nevertheless, this last criterion and the requirement of accessible (public) schoolyards to local residents address issues intertwined with the incorporation of (in)equality aspects into the climate resilience agenda. Both criteria are based on the idea that cities are not uniform landscapes and communities and therefore resilience practices need to consider disparities and different vulnerabilities (Vale 2014). In any case, our findings suggest that criteria related to climate justice or social equity were not the most prioritized in the school selection process, at least in Paris and Barcelona. Given the existing school green inequalities observed in cities like Barcelona (see Baró et al. 2021), we contend that the upscaling of these initiatives should consider the justice and equity dimensions more prominently. To this end, it is also important to understand why some schools might not be willing to engage in this type of projects (as apparently it was the case in Barcelona). In other countries, it has been observed that schools with lower resources cannot access schoolyard greening programmes because of the co-funding requirements established by the municipalities or the corresponding funding organization, including the extra costs that the maintenance of NBS usually entail (Jansson et al. 2019; Giezen and Pellerey 2021).

The three studied projects implemented a variety of NBS contributing to mitigate heat stress and stormwater runoff. However, it is also evident from the interviews and the project documents that climate nature-based solutions were also designed and implemented to offer multiple co-benefits related to educational, play, health or social cohesion opportunities (through the opening of green schoolyards to local residents). Besides, all projects also implemented hybrid (e.g. vegetated pergolas) and grey solutions (e.g. cool roofs) indicating that NBS and *classic* grey solutions can be synergetic rather than alternative options in this type of infrastructure transformations (see also Kabisch et al. 2017; Lin et al. 2021).

As suggested by different authors (Frantzeskaki and Kabisch 2016; Raymond et al. 2017), the design of NBS implementation processes needs to enable cross-sectoral partnerships and support openness, transparency and legitimacy of knowledge from citizens, practitioners and policy stakeholders. In the three analysed projects, this approach was addressed through the establishment of multidisciplinary teams (including the involvement of different City Council departments and interdisciplinary coordination profiles such as the Urban Resilience Officer in Paris) and the early engagement with multiple *end users*, including schoolchildren, school staff, parents and also urban residents more widely, throughout the different stages of the implementation process. In fact, the ambitious co-design strategy developed in all projects and applied in most pilot schools (from diagnosis of current state to evaluation of implemented measures) contributed to overcome several implementation challenges such as negative perceptions by some stakeholders, a problem frequently cited in other NBS projects (Collier et al. 2016). The exclusion of potential users, their knowledge and experience from the co-design process could negatively affect the effectiveness of the resulting adaptation measures (Muñoz-Erickson et al. 2017). This limitation is not unique in the case of co-designing NBS in school environments. Van der Ven et al. (2016) note that who should be engaged in the co-design process is an unresolved question for co-production in climate change adaptation in general. Institutional willingness for promoting inclusivity and equality in the co-production process is thus necessary (Wamsler 2017), as well as deep understanding of the stakeholders shaping place dynamics and uses to give voice to those who usually play the role of observers in the co-design processes (Ruiz-Mallén 2020).

The upscaling of the projects to more schools is planned as of 2021 in Paris and Barcelona. From our findings, strong political support at the highest municipal level and external funding (European UIA in this case) are key factors for the successful implementation and subsequent upscaling of these projects. Actually, a change in Madrid's government after 2019 municipal elections was clearly identified as a roadblock for the upscaling of Care in School Environments project. Upscaling efforts are also easier if framed within ambitious city-wide strategies related to urban greening, climate change adaptation or urban resilience (see also Fastenrath et al. 2020). The knowledge, scientific evidence and lessons learnt during the demonstration phase of the projects are also crucial for a successful mainstreaming phase. Besides, some NBS can produce additional or incremental co-benefits when scaled up. For example, despite schoolyards generally cover a small share of the total urban area, their transformation at the city level could generate substantial benefits in terms of habitat for species, pollination or stormwater runoff control similar to green roof upscaling efforts (Langemeyer et al. 2020). Moreover, schoolyards are generally not bound by urban planning regulations; hence, they are opportunity spaces for climate and social innovation/experimentation. The Covid-19 pandemic, despite it caused several additional implementation challenges, is also seen as an opportunity for school re-design and the creation of safe outdoor educational spaces.

This comparative exploration is not exhaustive, and it only includes the institutional perspective of project coordinators and available policy reports. Future research should include the experience and views of other involved stakeholders, especially *end users* such as schoolchildren and schoolteachers. These multiple social perceptions can be assessed through diverse methods, including Q method

(Buchel and Frantzeskaki 2015), narrative analysis (Gerstenberg and Hofmann 2016) or fuzzy cognitive mapping (Gray et al. 2015).

We contend that nature-based school transformations can be the spearhead for climate resilience and adaptation in cities due to the central role that schools play in urban life. Schools act as neighbourhood hubs, and their grounds can be re-purposed as spaces for social cohesion and intergenerational encounter around which to implement urban transformation capacities. Nature-based interventions in schools can also contribute towards climate justice as children will be the most affected by the impacts of climate change, yet those who have contributed the least to it. The exploration of these three pioneering projects suggests that school transformations are an opportunity for a radical urban social-ecological transition through the development of a strong climate and environmental awareness among those who will decide the fate of our cities during the second half of the twenty-first century.

Acknowledgements Authors acknowledge financial support from the European Research Council (project GREENLULUs; grant agreement ID: 678034), the EU's Horizon 2020 framework programme for research and innovation (project NATURVATION, grant agreement ID: 730243) and the JPI Urban Europe ERA-NET Cofund Urban Transformation Capacities call (project COOLSCHOOLS). This work also contributes to the "María de Maeztu" Programme for Units of Excellence of the Spanish Ministry of Science and Innovation (CEX2019-000940-M). We are also grateful to all the interviewees for accepting to participate in this study. I. Ruiz-Mallén acknowledges the financial support of the grant RYC-2015-17676 funded by MCIN/AEI/10.13039/501100011033 and by "ESF Investing in your future".

References

Akpinar A (2016) How is high school greenness related to students' restoration and health? Urban For Urban Green 16:1–8. https://doi.org/10.1016/j.ufug.2016.01.007

Amoly E, Dadvand P, Forns J et al (2014) Green and blue spaces and behavioral development in Barcelona schoolchildren: the BREATHE project. Environ Health Perspect 122:1351–1358. https://doi.org/10.1289/ehp.1408215

Antoniadis D, Katsoulas N, Papanastasiou DK (2020) Thermal environment of urban schoolyards: current and future design with respect to children's thermal comfort. Atmosphere (Basel) 11. https://doi.org/10.3390/atmos11111144

Barcelona City Council (2018) Barcelona Climate Plan 2018–2030. http://hdl.handle.net/11703/109218. Accessed 15 Oct 2020

Baró F, Camacho DA, Pérez Del Pulgar C et al (2021) School greening: right or privilege? Examining urban nature within and around primary schools through an equity lens. Landsc Urban Plan 208:104019. https://doi.org/10.1016/j.landurbplan.2020.104019

Bell AC, Dyment JE (2008) Grounds for health: the intersection of green school grounds and health-promoting schools. Environ Educ Res 14:77–90. https://doi.org/10.1080/13504620701843426

Broom C (2017) Exploring the relations between childhood experiences in nature and young adults' environmental attitudes and behaviours. Aust J Environ Educ 33:34–47. https://doi.org/10.1017/aee.2017.1

Browning MHEM, Rigolon A (2019) School green space and its impact on academic performance: a systematic literature review. Int J Environ Res Public Health 16. https://doi.org/10.3390/ijerph16030429

Buchel S, Frantzeskaki N (2015) Citizens' voice: a case study about perceived ecosystem services by urban park users in Rotterdam, the Netherlands. Ecosyst Serv 12:169–177. https://doi.org/10.1016/j.ecoser.2014.11.014
Chawla L, Keena K, Pevec I, Stanley E (2014) Green schoolyards as havens from stress and resources for resilience in childhood and adolescence. Health Place 28:1–13. https://doi.org/10.1016/j.healthplace.2014.03.001
Chawla L (2015) Benefits of nature contact for children. J Plan Lit 30:433–452. https://doi.org/10.1177/0885412215595441
Cohen-Shacham E, Andrade A, Dalton J et al (2019) Core principles for successfully implementing and upscaling nature-based solutions. Environ Sci Pol 98:20–29. https://doi.org/10.1016/j.envsci.2019.04.014
Collier MJ, Connop S, Foley K et al (2016) Urban transformation with TURAS open innovations; opportunities for transitioning through transdisciplinarity. Curr Opin Environ Sustain 22:57–62. https://doi.org/10.1016/j.cosust.2017.04.005
Dadvand P, Nieuwenhuijsen MJ, Esnaola M et al (2015) Green spaces and cognitive development in primary schoolchildren. Proc Natl Acad Sci 112:7937–7942. https://doi.org/10.1073/pnas.1503402112
de Coninck, H., A. Revi, M. Babiker, et al. (2018) Strengthening and implementing the global response. In: Global Warming of 1.5°C. An IPCC Special Report on the impacts of global warming of 1.5°C above pre-industrial levels and related global greenhouse gas emission pathways, in the context of strengthening the global response to the threat of climate change, sustainable development, and efforts to eradicate poverty. Cambridge University Press.
Demuzere M, Orru K, Heidrich O et al (2014) Mitigating and adapting to climate change: multi-functional and multi-scale assessment of green urban infrastructure. J Environ Manag 146:107–115. https://doi.org/10.1016/j.jenvman.2014.07.025
Equal Saree (ed) (2019) Inclusive school Playgrounds. A guide to diagnosis and intervention with a gender perspective. Pol·len ed. Equal Saree, Barcelona
Faber Taylor A, Kuo FE (Ming) (2011) Could exposure to everyday green spaces help treat ADHD? Evidence from children's play settings. Appl Psychol Heal Well Being 3:281–303. https://doi.org/10.1111/j.1758-0854.2011.01052.x
Fastenrath S, Bush J, Coenen L (2020) Scaling-up nature-based solutions. Lessons from the Living Melbourne strategy. Geoforum 116:63–72. https://doi.org/10.1016/j.geoforum.2020.07.011
Frantzeskaki N, Kabisch N (2016) Designing a knowledge co-production operating space for urban environmental governance—Lessons from Rotterdam, Netherlands and Berlin, Germany. Environ Sci Pol 62:90–98. https://doi.org/10.1016/j.envsci.2016.01.010
García-Serrano P, Urda-Peña L, Leal-Laredo P, de Blas M (2017a) Estudio MICOS: entornos escolares saludables. Bases para una estrategia de intervención de patios, accesos y entornos escolares http://madridsalud.es/wp-content/uploads/2016/04/MICOS-Regeneracion-urbana.pdf. Accessed 15 Nov 2020
García-Serrano P, Leal-Laredo P, Urda-Peña L (2017b) Guía de diseño de entornos escolares. Madrid Salud. City of Madrid. https://www.madridsalud.es/pdf/guia_diseno_entornos_escolares_opt.pdf. Accessed 15 Nov 2020
Gerstenberg T, Hofmann M (2016) Perception and preference of trees: a psychological contribution to tree species selection in urban areas. Urban For Urban Green 15:103–111. https://doi.org/10.1016/j.ufug.2015.12.004
Gidlow CJ, Ellis NJ (2011) Neighbourhood green space in deprived urban communities: issues and barriers to use. Local Environ 16:989–1002. https://doi.org/10.1080/13549839.2011.582861
Giezen M, Pellerey V (2021) Renaturing the city: factors contributing to upscaling green schoolyards in Amsterdam and The Hague. Urban For Urban Green 63:127190. https://doi.org/10.1016/j.ufug.2021.127190
Gray SA, Gray S, De Kok JL et al (2015) Using fuzzy cognitive mapping as a participatory approach to analyze change, preferred states, and perceived resilience of social-ecological systems. Ecol Soc 20

Jansson M, Vogel N, Fors H, Randrup TB (2019) The governance of landscape management: new approaches to urban open space development. Landsc Res 44:952–965. https://doi.org/10.1080/01426397.2018.1536199

Kabisch N, Korn H, Stadler J, Bonn A (eds) (2017) Nature-based solutions to climate change adaptation in urban areas: linkages between science, policy and practice. Springer, Cham

Kuo M, Klein SE, Browning HEM, Zaplatosch J (2021) Greening for academic achievement: prioritizing what to plant and where. Landsc Urban Plan 206:103962. https://doi.org/10.1016/J.LANDURBPLAN.2020.103962

Langemeyer J, Wedgwood D, McPhearson T et al (2020) Creating urban green infrastructure where it is needed – a spatial ecosystem service-based decision analysis of green roofs in Barcelona. Sci Total Environ 707:135487. https://doi.org/10.1016/J.SCITOTENV.2019.135487

Lin BB, Ossola A, Alberti M et al (2021) Integrating solutions to adapt cities for climate change. Lancet Planet Heal 5:e479–e486. https://doi.org/10.1016/S2542-5196(21)00135-2

Markevych I, Tiesler CMT, Fuertes E et al (2014) Access to urban green spaces and behavioural problems in children: results from the GINIplus and LISAplus studies. Environ Int 71:29–35. https://doi.org/10.1016/j.envint.2014.06.002

Madrid City Council (2016–2018) Plan MAD-RE (Madrid Recupera). https://www.madrid.es/portales/munimadrid/es/Inicio/Vivienda-y-urbanismo/Plan-MAD-RE/?vgnextfmt=default&vgnextchannel=097867c77d662610VgnVCM1000001d4a900aRCRD&vgnextoid=097867c77d662610VgnVCM1000001d4a900aRCRD. Accessed 15 Nov 2020

Madrid City Council (2017) Plan de Calidad de Aire de la Ciudad de Madrid y Cambio Climático (Plan A). https://www.madrid.es/portales/munimadrid/es/Inicio/Medio-ambiente/Publicaciones/Plan-de-Calidad-de-aire-de-la-ciudad-de-Madrid-y-Cambio-Climatico-PLAN-A-/?vgnextfmt=default&vgnextoid=2b809df12834b510VgnVCM1000001d4a900aRCRD&vgnextchannel=f6ff79ed268fe410VgnVCM1000000b205a0aRCRD. Accessed 15 Nov 2020

Madrid City Council (2017–2021) Plan Madrid Ciudad de los Cuidados. https://transparencia.madrid.es/portales/transparencia/es/Organizacion/Planes-y-memorias/Planes/Plan-Madrid-Ciudad-de-los-Cuidados-2017-2021/?vgnextfmt=default&vgnextoid=19bb1a73a780061 0VgnVCM2000001f4a900aRCRD&vgnextchannel=d869508929a56510VgnVCM1000008a4a900aRCRD. Accessed 15 Nov 2020

Muñoz-Erickson TA, Miller CA, Miller TR (2017) How cities think: knowledge co-production for urban sustainability and resilience. Forests 8:203. https://doi.org/10.3390/f8060203

Nesshöver C, Assmuth T, Irvine KN et al (2017) The science, policy and practice of nature-based solutions: an interdisciplinary perspective. Sci Total Environ 579:1215–1227. https://doi.org/10.1016/j.scitotenv.2016.11.106

Pain, R, Whitman, G, Milledge, D, Trust, LR (2012) Participatory action research toolkit: an introduction to using PAR as an approach to learning, research and action. Durham University, RELU, Lune Rivers Trust

Paris City Council (2018) Paris Reslience Strategy. https://www.paris.fr/pages/paris-resiliente-4264/. Accessed 15 Oct 2020

Pérez-del-Pulgar C, Anguelovski I, Cole HVS et al (2021) The relationship between residential proximity to outdoor play spaces and children's mental and behavioral health: the importance of neighborhood socio-economic characteristics. Environ Res 200:111326. https://doi.org/10.1016/j.envres.2021.111326

Raymond CM, Frantzeskaki N, Kabisch N et al (2017) A framework for assessing and implementing the co-benefits of nature-based solutions in urban areas. Environ Sci Pol 77:15–24. https://doi.org/10.1016/j.envsci.2017.07.008

Rosa CD, Profice CC, Collado S (2018) Nature experiences and adults' self-reported pro-environmental behaviors: the role of connectedness to nature and childhood nature experiences. Front Psychol 9:1055. https://doi.org/10.3389/fpsyg.2018.01055

Ruiz-Mallén I (2020) Co-production and resilient cities to climate change. In: Narez J, Bole D (eds) Participatory research and planning in practice, The urban book series. Springer, Cham, pp 1–11. ISBN: 978-3-030-28013-0

Shape America (Society of Health and Physical Educators) (2016) Guide for recess policy. Reston, VA: Society of Health and Physical Educators. https://www.shapeamerica.org//advocacy/upload/Guide-for-Recess-Policy.pdf. Accessed 15 Dec 2020

Slater SJ, Nicholson L, Chriqui J et al (2012) The impact of state Laws and District policies on physical education and recess practices in a nationally representative sample of US public elementary schools. Arch Pediatr Adolesc Med 166:311–316. https://doi.org/10.1001/archpediatrics.2011.1133

Vale L (2014) The politics of resilient cities: whose resilience and whose city? Build Res Inf 42(2):191–201

van der Ven FHM, Snep RPH, Koole S, Brolsma R, van der Brugge R, Spijker J, Vergroesen T (2016) Adaptation planning support toolbox: measurable performance information based tools for co-creation of resilient, eco-system-based urban plans with urban designers, decision-makers and stakeholders. Environ Sci Pol 66:427–436. https://doi.org/10.1016/j.envsci.2016.06.010

van Dijk-Wesselius JE, Maas J, Hovinga D et al (2018) The impact of greening schoolyards on the appreciation, and physical, cognitive and social-emotional well-being of schoolchildren: a prospective intervention study. Landsc Urban Plan 180:15–26. https://doi.org/10.1016/j.landurbplan.2018.08.003

van Dijk-Wesselius JE, van den Berg AE, Maas J, Hovinga D (2020) Green schoolyards as outdoor learning environments: barriers and solutions as experienced by primary school teachers. Front Psychol 10:2919. https://doi.org/10.3389/fpsyg.2019.02919

Vetter R (2020) Cooling schools: experiences from C40's Cool Cities Network. https://www.c40knowledgehub.org/s/article/Cooling-schools-Experiences-from-C40s-Cool-Cities-Network?language=en_US. Accessed 15 Dec 2020

Wamsler C (2017) Stakeholder involvement in strategic adaptation planning: transdisciplinarity and co-production at stake? Environ Sci Pol 75:148–157. https://doi.org/10.1016/j.envsci.2017.03.016

Wild T, Bulkeley H, Naumann S, et al (2020) Nature-based solutions: State of the art in EU-funded projects. https://op.europa.eu/s/o1zq. Accessed 15 Dec 2020

Chapter 7
Social-Ecological Transformation to Coexist with Wildfire: Reflecting on 18 Years of Participatory Wildfire Governance

Iago Otero

Abstract The risk of devastating wildfires – exacerbated by climate change – poses a threat to urban areas worldwide. There is a pressing need to strengthen societal efforts to coexist with this perturbation by creating resilient social-ecological systems. To enable this, a significant social-ecological transformation of wildfire-prone regions seems to be required. Here, I reflect on my experience in three projects of participatory wildfire governance conducted in the Barcelona Metropolitan Region (Spain) during the last 18 years. The goal is to learn from the experience and to sketch new transformative options to coexist with wildfire. After a literature review on the links between resilience to wildfire, adaptation and transformation, I analyse these projects with regard to their achievements, challenges and potential new transformative avenues. The analysis shows the crucial role that a locally rooted civil society can have when it is able to network with key agencies and actors over the long term. It also shows the importance of developing integrative wildfire planning networks where different ecosystem services and values are considered in successive phases of public deliberation between actors, citizens and wildfire managers. It is concluded that deepening the transdisciplinary content of participatory wildfire governance can increase its transformative potential.

Keywords Wildfire · Transformative adaptation · Participatory wildfire governance · Resilience

---

I. Otero (✉)
Interdisciplinary Centre for Mountain Research, University of Lausanne, Lausanne, Switzerland
e-mail: iago.otero@unil.ch

I. Ruiz-Mallén et al. (eds.), *Urban Resilience to the Climate Emergency*, The Urban Book Series, https://doi.org/10.1007/978-3-031-07301-4_7

7.1 Introduction

Wildfires are increasingly a matter of urban concern. I am now watching a picture of the Golden Gate Bridge of San Francisco taken on 9 September 2020. The lights of the bridge flickered amidst a reddish atmosphere that covered the entire city with smoke. The state of California was going through an outstanding wildfire episode that lasted for weeks. In August of that year, the daily wildfire radiative power had massively exceeded the average values of the period 2003–2019 (Fig. 7.1). *Alarming, but unsurprising*, said experts quoted by the press. Indeed, between 1972 and 2018, annual burned area in California has increased fivefold, and summer fire extent up to eightfold, most likely due to increased aridity caused by climate change (Williams et al. 2019). Wildfire episodes affecting hundreds of thousands of hectares and producing casualties and infrastructural damage are thus expected to become more frequent. At the heart of the problem, there is the expansion of the so-called Wildland-Urban Interface (WUI), i.e. residential developments spreading across highly flammable forestland.

In Europe, the risk of increasingly devastating wildfires in the WUI is considered one of the main challenges faced by fire departments (Castellnou et al. 2019). In the last decades, Mediterranean forests of Europe have witnessed an increase in the

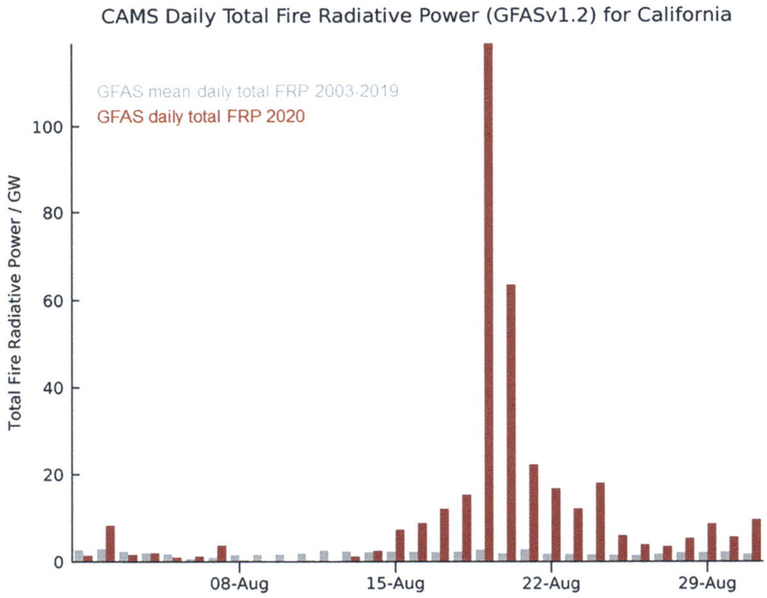

Fig. 7.1 Daily total Fire Radiative Power (FRP) for California throughout August 2020 (red) and the 2003–2019 daily average for the month (grey). The GFAS, meaning Global Fire Assimilation System, assimilates FRP observations from satellite-based sensors to produce daily estimates of emissions from wildfires and biomass burning. (Source: Copernicus Atmosphere Monitoring Service, European Centre for Medium-Range Weather Forecasts (2020))

length of fire weather seasons due to a changing climate (Jolly et al. 2015). Regarding the number of large wildfires (above 500 ha), even if no increasing trend is detected, these events have been catastrophic for entire regions due to human casualties and socioeconomic losses (San-Miguel-Ayanz et al. 2013). Projections for the coming decades suggest a sharp increase in the number of days with meteorological conditions conducive to extreme wildfires (Bowman et al. 2017). Summer burnt area is thus projected to increase between 40% and 100% (Turco et al. 2018; see also Moritz et al. 2012), even if a considerable uncertainty remains in the models used (Brotons and Duane 2019). More northern European latitudes have recently experienced unprecedented wildfire events too, turning the wildfire challenge into a continent-wide matter of concern much beyond the Mediterranean (San-Miguel-Ayanz et al. 2018; Castellnou et al. 2019; see also Müller et al. 2020).

The situation is similar in other regions of the world, where extreme wildfire activity has increased or is expected to increase, including some regions of the UK (Albertson et al. 2010), some regions of Australia (O'Neill and Handmer 2012) and the Western USA (McKenzie et al. 2004; Westerling et al. 2006). Whereas fire activity is expected to increase in some areas of the planet and to decrease in others (Krawchuk et al. 2009; Moritz et al. 2012), globally the length of fire seasons has increased (Jolly et al. 2015), and fire seasons are expected to be more severe (Flannigan et al. 2013). Actually, a new global pattern of extreme wildfire events is observed from 2016, with erratic behaviours and unprecedented speed, intensity and extension (Duane et al. 2021). This is, for example, the case of the 2017 wildfires in Portugal, which reached propagation rates of up to 14,000 ha/h and caused 117 casualties due to complex firestorms that are just starting to be understood by wildfire managers and scientists (Castellnou 2020; San-Miguel-Ayanz et al. 2020; Duane et al. 2021).

Many of the world's fire-prone regions are densely populated and have high social, economic and ecological values at stake (Moritz et al. 2014; Doerr and Santín 2016; Otero et al. 2018). Thus, there is a pressing need to strengthen societal efforts to coexist with wildfire in their urban and metropolitan areas. Coexisting with wildfire is here understood in close connection with both adaptation and resilience. Adaptation can be broadly defined as a social process aimed at minimizing the adverse impacts of expected wildfires via the creation of resilient social-ecological systems – including urbanized environments – able to withstand this perturbation without major losses in social and ecological functions (Prior and Eriksen 2013; González-Hidalgo et al. 2014; Abrams et al. 2015; Fischer and Jasny 2017). The focus of this chapter is on *transformative* adaptation, highlighting that resilience to wildfires requires a radical social-ecological transformation including for instance the political economy of land use, institutional cultures and nature's values (Otero and Nielsen 2017). Participatory governance is a key element of such necessary transformations (Otero et al. 2018). However, the literature has not yet provided studies assessing the transformative effect of participatory governance efforts, especially in the mid-term (decades).

In this chapter I look back to my experience in three projects of participatory wildfire governance conducted in the last 18 years, which I treat as case studies. The projects are located in the outer ring of the Barcelona Metropolitan Region, in Spain. The goal of this self-reflexive approach (Lang et al. 2012; Popoveniuc 2014; Schneider et al. 2019; Otero et al. 2020a) is to learn from the accumulated experience and to sketch new transformation options to coexist with wildfire. To do so, I (i) reviewed the main publications that emerged from these projects, (ii) drew on my personal memories (thus, the analysis has a subjective component) and (iii) updated on the projects' development through informal conversations and email exchanges with the colleagues involved in them. In all cases, the analysis focused on the transformative potential of the projects, in hindsight. The chapter is structured as follows. Section 7.2 provides a synthesis of the literature on resilience and adaptation to wildfire, with the emerging focus on transformation. Section 7.3 provides the context of the case studies, and Sect. 7.4 describes them in detail. Section 7.5 assesses their transformative potential, and Sect. 7.6 closes with some conclusions. I use *the author* to refer to myself.

7.2 Coexisting with Wildfire Requires a Radical Social-Ecological Transformation

7.2.1 Proposals to Adapt and Build Resilience to Wildfire

There is a growing literature on how to better coexist with wildfire. The focus of this literature is often what kind of social relationships facilitate resilience and adaptation. A review of social science research on wildfire found that social relationships in wildfire-prone areas can increase wildfire preparedness by building a sense of community that facilitates the exchange of crucial information (McCaffrey 2015). Similarly, Prior and Eriksen (2013) showed that community cohesion supports the adoption of protective measures (such as the reduction of combustible materials around houses) that contribute to adaptive capacity and resilience. This is so because people draw on social cohesion to get the support and resources necessary to undertake such measures (Prior and Eriksen 2013). Gender relations also matter. Eriksen (2014), for example, concluded that the ability of all family members to contribute to wildfire preparedness is crucial to support resilience strategies, challenging gendered patterns of vulnerability.

Beyond the community level, the importance of social relationships for adaptive capacity and resilience is also recognized in the literature about the so-called participatory wildfire planning networks. These networks are made of public agencies, actors and citizens, and they are being co-developed in different wildfire-prone countries of the world (Moore et al. 2002; Chapin et al. 2008; Butler and Goldstein 2010; Everett and Fuller 2011; Almstedt and Reed 2013; Plana et al. 2015; Gazzard et al. 2016; Otero et al. 2018, 2020b). Such co-production is driven both by a normative motivation to democratize wildfire management (Morehouse and O'Brien 2008; Otero et al. 2018) and by the recognition that the complexity of the wildfire problem requires new governance arrangements operating at multiple scales (Butler and Goldstein 2010; Ager et al. 2015; Fischer et al. 2016). Indeed, community wildfire resilience can be enhanced by adaptive governance mediated by institutions at multiple scales, as it opens social opportunities to learn from and adapt to wildfire (Abrams et al. 2015; Otero et al. 2018). Collaboration across scales and agencies, as well as community involvement, are considered key factors for effectively planning wildfire resilience schemes (Plana et al. 2015; Tàbara et al. 2003), not least in urban and metropolitan settings. A crucial inter-agency link highlighted in the literature is that between wildfire risk management and land planning. For example, Moritz et al. (2014) argued that governments and agencies should restrict urban development in the most wildfire-prone locations (see also Plana 2011). Badia et al. (2019) in turn suggested the need to adapt wildfire prevention schemes to the different types of existing residential developments. In urban and metropolitan settings, participatory wildfire planning networks have to integrate a wide diversity of interests at stake while deliberating about alternative planning and policy options. To facilitate this endeavour, different tools have been used with varying degrees of success, like social-ecological models of wildfire risk (Fischer et al. 2016; Spies et al. 2014) or participatory multi-criteria evaluation methods (Otero et al. 2020b; Gamboa et al. under review).

Regardless of the importance of building the right social and institutional relationships, some wildfire practitioners and researchers have long emphasized that a better coexistence with wildfire requires fire restoration. Fire restoration refers to the reintroduction of (wild)fire in ecosystems after decades of suppression policies, which are considered to be one of the underlying causes of increased risk as they enable fuel build-up (Donovan and Brown 2007; Calkin et al. 2015; Otero and Nielsen 2017). Thus, authors stressed the need for incentives that encourage fire managers to consider the beneficial effects of allowing wildfires to burn under certain circumstances rather than suppressing them all as quick as possible (Donovan and Brown 2007; North et al. 2015; Thompson et al. 2015). This requires fundamental organizational changes in how the fire departments think about, plan for and respond to wildfires (Thompson et al. 2018; Castellnou et al. 2019). The traditional suppression rationale of the wildfire management system can also be overcome by developing multiscale participatory planning networks that help overcome institutional barriers to fire restoration (Butler and Goldstein 2010).

Letting some wildfires burn under previously planned conditions could be a cost-effective fuel reduction strategy (Regos et al. 2014; Domènech et al. 2018; Oliveres 2021) even if considerable challenges remain regarding the social acceptance of this measure, especially in densely populated urban areas (Otero and Nielsen 2017). A softer, potentially more easily acceptable fire restoration measure is prescribed burning, consisting in the planned use of fire under specific environmental conditions for fuel load reduction. Studies showed that prescribed burning can reduce wildfire intensity under climate change, especially if combined with other treatments like thinning and when adapted to the spatial pattern of past wildfires (Piñol et al. 2005, 2007; Piqué and Domènech 2018; Duane et al. 2019). Whereas increased wildfire activity due to human influence can threaten species with extinction, fire restoration has the potential to benefit biodiversity if it is tailored to the requirements of particular ecosystems and species (Kelly and Brotons 2017; Kelly et al. 2020).

Finally, there are a number of other proposals to build wildfire-resilient social-ecological systems that are likely to be complementary with fire restoration attempts (Pais et al. 2020; Campos et al. 2021). These include fuel management through activities like extensive grazing (Otero 2011a; Varela et al. 2018), biomass extraction for energy production (Regos et al. 2016), the creation of new cropland or the recovery of old cropland in current forests (Aquilué et al. 2020) and the integration of wildfire risk into forestry practices (e.g. developing new management practices aimed at achieving enough separation between understory and overstory in order to prevent crown fires, Piqué et al. 2011). All together, these proposals entail a recovery (or remaking) of the old rural mosaic, which can be suitable to reduce climate change-induced increases in wildfire intensity and burnt area (Loepfe et al. 2010, 2012).

7.2.2 The Need for Transformative Approaches in Wildfire Governance

Coexisting with wildfire in a rapidly changing Earth seems to require a number of interlinked social, institutional, cultural and ecological changes: cohesive communities, inter-agency collaboration, citizen involvement, shift from fossil to biomass energy and accepting that our landscape may need to burn if it is to become resilient. Altogether, these changes hint at a profound transformation of the current social-ecological relationships (Howitt 2014). Indeed, O'Neill and Handmer (2012) pointed out the need for transformative adaptation, i.e. an adaptation which is concerned with the root causes of wildfire vulnerability and that calls for deep shifts in people's relationships with their environment.

Transformative adaptation for example includes a substantive cultural change towards accepting that wildfires are intrinsic part of some ecosystems, especially under climate change. To enable such change, the entrenched habit of attributing blame for wildfires to certain people (e.g. a mistake in the fire department's operations) should cease, especially in parliamentary debates (González-Hidalgo et al. 2013, 2014). The latter should instead endorse constructive approaches that promote learning while addressing the root causes of wildfire risk (González-Hidalgo et al. 2014). Such causes include rural exodus and the abandonment of the countryside, forest expansion, urbanization, fossil fuel-based food and energy systems and a culture aspiring to control nature (Otero and Nielsen 2017). To address these causes and build resilience to wildfire, authors stressed the need to create sites of *healthy contestation* that acknowledge the political dimension of the choices that need to be made (Buizer and Kurz 2016). In these sites, the political economy of land use underlying wildfire risk can be opened to debate and antagonism channelled into transformative avenues (González-Hidalgo et al. 2014; Otero et al. 2018). Similarly, the creation of *political communities* able to deal with conflicting values (e.g. wildfire security vs. scenic beauty in the WUI) while shaping less flammable landscapes was suggested as a way forward in this direction (Moore et al. 2002; Morehouse and O'Brien 2008). In theory, these communities have the potential to redefine agency over landscape, i.e. the power to actively shape it, when co-designing wildfire resilience schemes (Otero et al. 2018).

However, all these changes pertaining to transformative adaptation are not easy to implement. The wildfire-prone regions of the Global North are warming, urbanizing and afforesting at fast rates, and the efforts underway to coexist with wildfire are not able to counteract these underlying trends (Otero and Nielsen 2017). Thus, more research is needed on the transformative processes behind these efforts. In particular, we need more and better knowledge about their obstacles and their catalysers.

7.3 The Barcelona Metropolitan Region (Spain): A Vulnerable Landscape

The three case studies that will be analysed are located in the Barcelona Metropolitan Region (BMR) (Autonomous Community of Catalonia, Spain). The BMR hosts 5.2 million people, which represents two thirds of the Catalan population (IERMB 2021), in a territory that amounts to ca. 10% of the Autonomous Community. The BMR is structured in a compact centre (the municipality of Barcelona), a dense first metropolitan ring (composed of 32 municipalities), an extensive second metropolitan ring (124 municipalities) and 7 urban sub-centres of between 50,000 and more than 200,000 inhabitants (Catalán et al. 2008). Two mountain ranges parallel to the coastline dominate the BMR: the coastal range (with Garraf, Collserola, Serralada de Marina, Serralada Litoral and Montnegre-Corredor mountains) and the pre-coastal range (with Montserrat, Sant Llorenç del Munt-l'Obac, Cingles de Bertí and Montseny mountains). Contrasting topography (elevation ranges from 0 to 1700 m a.s.l.) and climate (with gradients NE–SW from moist to dry and SE–NW from less to more continental) brings about a high environmental variability, which together with a long-term imprint of human activities shaped remarkably heterogeneous landscapes. Coastal and pre-coastal ranges are mostly covered by continuous forest and shrub, whereas lowland areas are dominated by croplands, forest-crop mosaics and urban areas (Marull et al. 2010). The BMR accounts for more than 40 habitats of European significance, including many species of fauna and flora that are either endangered or threatened with extinction (Marull et al. 2007). Most of the natural protected areas of the BMR are located in mountain areas (Otero and Boada 2007).

Looking back to history allows to better understand the current context of the case studies regarding wildfire risk. On the one hand, the unfolding of urbanization and industrialization, together with globalization trends, led to a diminishing profitability of forestry (Boada and Zahonero 1998; Otero et al. 2008). Low productivity and the mountainous relief made these forests not competitive enough, as it happened in most of the Catalan forests (Rojas 1999). As a result, they were gradually left unmanaged leading to vegetation regrowth and fuel accumulation. On the other hand, cropland abandonment occurred as farmers shifted to non-farm jobs and/or moved to towns (e.g. Otero et al. 2013), triggering afforestation processes that added up potentially burnable fuel. In parallel, urbanization often occurred in forest areas or in areas undergoing afforestation (e.g. Otero et al. 2011), creating a WUI which is highly vulnerable to wildfire. Between 1956 and 2000, 14% of the BMR experienced urbanization, 10% afforestation and 7% farmland abandonment (Başnou et al. 2013). Large wildfires become specially threatening under extreme weather conditions, which are not only relatively common in the Mediterranean climate of the BMR but also increasing with climate change (Brotons et al. 2013; Duane and Brotons 2018; Duane et al. 2021). Under current vegetation and climatic conditions, experts consider that the main mountain ranges of the BMR (Collserola, Garraf, Montnegre-Corredor) have the potential to burn entirely in less than one day (Pahisa 2021).

7.4 Participatory Wildfire Governance: Three Case Studies

The three case studies analysed in this chapter are attempts to build resilience to wildfire by participatory adaptive governance, where the production of scientific knowledge blurs with the co-design and implementation of management actions on the ground by different actors. Table 7.1 provides a detailed description of each case study, including the author's changing role in them. The approach used in the case studies was informed by three main concepts or forms of research and practice, which partly overlap: transdisciplinarity, action research and adaptive governance. Transdisciplinarity can be defined as a research procedure that integrates scientists from different disciplines, non-academic actors and citizens in the co-production of knowledge that is intended to address complex societal and environmental problems (Darbellay et al. 2014; Lang et al. 2012). Even if a direct link between transdisciplinarity and real societal transformations still lacks empirical support (Schäfer et al. 2020), scholars working in this domain implicitly consider that transdisciplinarity has a transformative potential (Otero et al. 2020a). Action research refers to *a participatory process concerned with developing practical knowing in the pursuit of worthwhile human purposes* (Reason and Bradbury 2008: 4). In this process, insider or outsider scientists work together with local communities to address a particularly problematic situation by trying to initiate desirable social-ecological transformations (Otero 2011b). At the same time, the knowledge produced through action research can be transferred beyond its immediate context so that others can see its application to their contexts (Herr and Anderson 2005), for instance, via peer-reviewed scientific literature. In such transference, this knowledge may be

Table 7.1 Basic description of case studies

Description	Case study #1: Municipality of Matadepera	Case study #2: Region Montseny-Montnegre-Corredor	Case study #3: Montseny Biosphere Reserve
Social-ecological context	Area: 25.4 km^2. Population: 9496 (2020). Elevation range: ca. 400–1100 m.a.s.l. Located on the hillside of the pre-coastal mountain range of Sant Llorenç del Munt-l'Obac, outer ring of the Barcelona Metropolitan Region 61% belongs to the Natural Park Sant Llorenç del Munt-l'Obac Settlement composed of compact town centre and wealthy low-density housing	Area: 606.0 km^2. Elevation range: ca. 0–1700 m.a.s.l. It includes the coast, the coastal mountain range of Montnegre-Corredor, part of the pre-coastal mountain range of Montseny, and a plain lying between the two mountain ranges Both mountain ranges are protected by natural parks, and Montseny is also a UNESCO Biosphere Reserve Located in the Barcelona Metropolitan Region. The plain and the coast host industrial activities, strategic transport infrastructures and touristic towns Settlement structured in towns, low-density housing and scattered farmhouses	Area: 501.7 km^2. Population: 51,573 (2014). Elevational gradient (max. Altitude is 1700 m.a.s.l.) allows the existence of three biogeographic strata (Mediterranean, Euro-Siberian and Alpine). This, together with a long history of human use, is the origin of a rich biodiversity. It includes 18 municipalities covering the pre-coastal mountain range of Montseny and surrounding plains. Besides the Biosphere Reserve, these municipalities are also partly protected by the Montseny Natural Park It has a core area (conservation function), a buffer area (function of making compatible conservation and human activities) and a transition area (development function)
What was done	Drafting and implementation of a participatory and adaptive land management scheme aimed at building resilience to large wildfires	Set-up of a pilot participatory network to democratize the Fire Department's wildfire prevention and suppression planning and build resilience to large wildfires	Improvement of the pilot participatory network and the method of case study #2 project, with the goal of democratizing the Fire Department's wildfire prevention and suppression planning and building resilience to large wildfires (see case study #2)

(continued)

Table 7.1 (continued)

Description	Case study #1: Municipality of Matadepera	Case study #2: Region Montseny-Montnegre-Corredor	Case study #3: Montseny Biosphere Reserve
By whom	Group of Forest Defence (*Agrupació de Defensa Forestal, ADF*), a municipal association composed of representatives of the Town Council, forest owners and volunteers but managed by volunteers In collaboration with the Local Environmental Council, municipal agencies, the Autonomous University of Barcelona, the Natural Park and other actors Funded by the Matadepera Town Council, the Catalan Department of the Environment, the Natural Park and private foundations	Led by a post-doctoral researcher studying wildfires and social-ecological transformations (Humboldt University of Berlin), the head of the Fire Department's expert group on wildfires (*Grup de Recolzament d'Actuacions Forestals, GRAF*) and a social activist working to empower citizens in democratic politics In collaboration with public agencies in charge of wildfire prevention and suppression, natural parks, town councils, associations of forest landowners, Groups of Forest Defence (*Agrupacions de Defensa Forestal, ADF*), museums and citizens Funded by the German Excellence Initiative through Humboldt University of Berlin	Led by a post-doctoral researcher with expertise in multi-criteria evaluation (Autonomous University of Barcelona) and a researcher with expertise in wildfires and social-ecological transformations (University of Lausanne). The project team included the *Pau Costa Foundation* (network of expertise in wildfire management and ecology) and *El Risell* (cooperative dealing with sustainability and social participation). It also counted on the advice from the Fire Department's expert group on wildfires (*Grup de Recolzament d'Actuacions Forestals, GRAF*) Implemented in collaboration with public agencies in charge of wildfire prevention and suppression, natural parks, town councils, an association of forest landowners, Groups of Forest Defence (*Agrupacions de Defensa Forestal, ADF*), museums and other actors Funded by *Fundación Biodiversidad* of the Spanish Ministry for Ecological Transition

How it was done	Delimitation of Strategic Management Areas (SMA; wildfire-friendly landscape structures that, if implemented, facilitate suppression and limit wildfire size). This delimitation was done by the Fire Department's expert group on wildfires (*Grup de Recolzament d'Actuacions Forestals, GRAF*) Design of a land management scheme with interventions to reduce fuel load in SMA: maintenance of cropland, thinning/pruning of forests and extensive grazing (ca. 140 ha) Approval of land management scheme by Town Council Implementation of interventions through formal and informal agreements with: farmers, forest owners, forest workers, shepherds and other actors Monitoring of interventions' results through observation and (short-term) scientific research Attempt to execute prescribed burns Public outreach on wildfire risk and mitigation measures through publications, media, talks, seminars, social networks etc. Accompanied by the maintenance of forest tracks and water infrastructures and by patrolling shifts during high-risk season (integrated in municipal and regional wildfire prevention schemes) Accompanied by actions to activate the social memory of the rural past (notably through historical research and outreach) and to protect local landscape and rural heritage from urbanization	Identification of expected wildfires, delimitation of wildfire containment polygons (areas used to assess wildfire potential spread) and delimitation of Strategic Management Areas (SMA; wildfire-friendly landscape structures that, if implemented, would facilitate suppression and limit wildfire size) Mapping alternative land management projects Involving regional actors (wildfire governance and others like museums) and co-designing a method to value the landscape of the wildfire containment polygons. The method consisted in gathering actors' input per categories of values (biodiversity, socioeconomic activities etc.) Applying the valuation method in four selected municipalities of the Montseny mountain range Organizing exhibitions in these municipalities in collaboration with municipal councils and *ADF*. The polygons' values were shown in posters, and citizens were invited to rank the polygons Planning a wildfire prevention and suppression strategy based on the priorities expressed in the exhibitions. Such strategy consisted in implementing those SMA protecting the most valued polygons Discussing and approving the strategy with the actors and citizens that participated	Characterization of the wildfire regime: study of wildfire types that affected the region, selection of a reference type (convection wildfire with wind), assessment of expected trends due to climate change and delimitation of wildfire containment polygons (areas used to assess wildfire potential spread) Participatory mapping of landscape values: data collection, definition of value categories and development of raster maps assigning a score to each pixel according to the values contained Participatory prioritization of wildfire containment polygons: selection and application of a multi-criteria evaluation method to rank polygons according to the quantity and diversity of values contained Planning a wildfire prevention and suppression strategy to protect the priority polygons. Such strategy consisted in delimiting and implementing Strategic Management Areas (SMA; wildfire-friendly landscape structures that, if implemented, would facilitate suppression and limit wildfire size) protecting the most valued polygons Discussing the strategy with the actors that participated

(continued)

Table 7.1 (continued)

Description	Case study #1: Municipality of Matadepera	Case study #2: Region Montseny-Montnegre-Corredor	Case study #3: Montseny Biosphere Reserve
Period	2003 to present	2014–2018	2018–2021
Author's involvement	2003–2013: co-leader as president of the Local Environmental Council (2003–2008) and president of *ADF* (2008–2013). Insider (living in the area) 2013–2021: follow-up of activities through communication with colleagues from *ADF*	2014–2016: leader as post-doctoral researcher of Humboldt University of Berlin. Partly insider (living in the area during fieldwork) and partly outsider (living in Berlin) 2017–2018: maintaining contact with the network for dissemination purposes, publication of results and preparation of extended phase (see case study #3)	2018–2019: co-leader as principal investigator of research project (working in the University of Lausanne). Outsider (living in Switzerland) 2019–2021: follow-up of activities through communication with the other co-leader

Source: (#1) Farriol et al. (2007), Otero (2010: 153–156), Otero (2011a). (#2) Otero et al. (2018). (#3) Otero et al. (2020a, b), Gamboa et al. (under review)

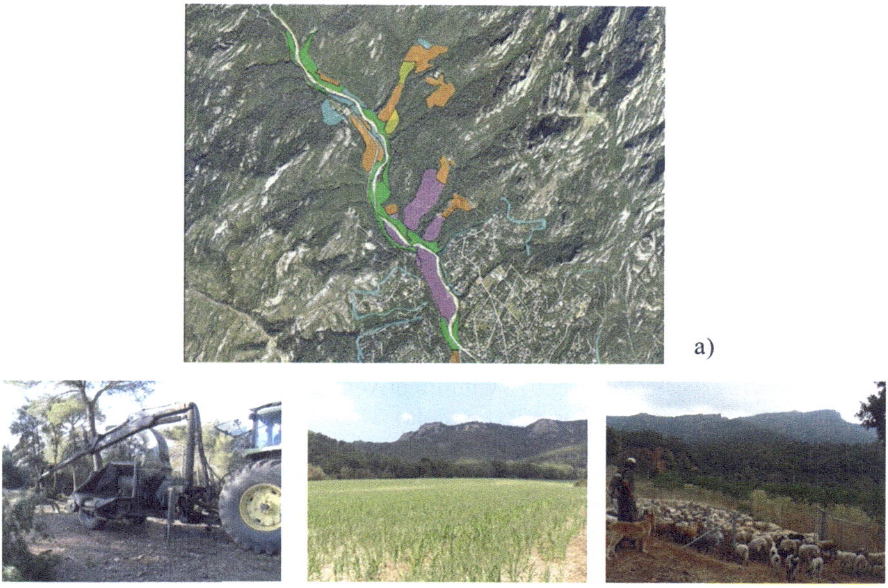

a)

b) c) d)

Fig. 7.2 Case study #1: Land management scheme to build resilience to wildfires by Matadepera's Group of Forest Defence (*Agrupació de Defensa Forestal, ADF*). (**a**) Different actions were planned to reduce fuel load in Strategic Management Areas (SMA). Example of the SMA along the main water stream. Orange: active cropland to be maintained. Yellow: abandoned cropland to be recovered. Pink: forest in flat terrain, to be thinned. Green: forest in steep terrain, to be thinned. Blue: low fuel strips around residential areas. (**b**) Forest thinning, consisting in felling a selection of trees and eventually cutting the understorey. (**c**) Cropland maintained by a family of farmers living in one of the farmhouses of Sant Llorenç del Munt, integrated within *ADF*'s management scheme. (**d**) First flock (sheep and goats) that came to Matadepera hired by *ADF*, with Sant Llorenç del Munt in the back. (Source: *ADF* Matadepera (**a**, **b**, **d**); I. Otero (**c**))

strengthened and then fed back to the context from which it emerged (Otero 2010). Finally, adaptive governance is a form of ecosystem management that connects individuals, organizations, agencies and institutions at multiple scales and organizational levels in order to build social-ecological resilience (Cash et al. 2006; Olsson et al. 2007). For these multi-level structures, a crucial challenge is to learn how to be prepared for disturbance and change. Actually, it is considered that social-ecological systems may use disturbances as opportunities to transform themselves into more desirable and resilient states (Folke et al. 2005).

Case study #1 is located in the municipality of Matadepera (9496 inhabitants), on the hillside of the mountain range of Sant Llorenç del Munt-l'Obac, which is protected by a natural park. Matadepera is a wealthy residential suburb located in the second ring of the BMR, neighbouring the cities of Terrassa and Sabadell, two metropolitan sub-centres of more than 200,000 inhabitants each. It has an extensive low-density WUI spanning between the historic centre and the natural park. Out of the three case studies analysed, it is by far the one that has been ongoing for the longest period. After a large wildfire burned 4600 ha in some neighbouring municipalities in 2003, the volunteer-led Group of Forest Defence (*Agrupació de Defensa Forestal, ADF*) of Matadepera started drafting a land management scheme aimed at building resilience to large wildfires (ADF Matadepera 2005). Back then, the author was living in the municipality and became one of the leaders of this initiative, notably as president of *ADF*. Essentially, the scheme consists in fuel reduction treatments in Strategic Management Areas (SMA) through forest thinning, cropland maintenance or extensive grazing (Fig. 7.2). SMA are areas where wildfires have the potential to decrease in intensity if a fuel reduction treatment has been implemented in advance and therefore help limit wildfire size and impacts. In Matadepera several SMA were planned along the main water stream. These SMA would allow fire brigades to contain a convective wildfire driven by west winds before reaching Sant Llorenç del Munt main summit (La Mola). These SMA were previously identified by the wildfire specialist group of the Catalan Fire Department (*Grup de Recolzament d'Actuacions Forestals, GRAF*) in collaboration with a forest engineer belonging to *ADF*, who drafted the scheme. The scheme was approved by the Town Council in 2005, and the first big interventions were implemented in 2007. The process is still active as of 2021.

Implementing these fuel reduction treatments required reaching written and informal agreements between varied institutions and actors. For example, thinning occurring within the borders of the natural park required the authorization of its management office. As most of the thinning occurred in private forests, the authorization of the forest owners was also necessary. A written agreement signed between *ADF* and the municipality's largest forest owner enabled the use of her farmyard to shelter flocks of sheep and/or goats that were meant to graze in the thinned forests to prevent shrub encroachment (Fig. 7.2d). Yearly written agreements and contracts between *ADF*, the Town Council and the shepherds regulated the financial and logistic support offered to the latter for grazing in these forests. Agricultural works like ploughing and sowing (notably to produce fodder for the flock) were entrusted to farmers. All these agreements required a sustained effort by *ADF* of networking, trust building and mediation between actors whose varied interests had to be made compatible for a common goal (see details of forestry interventions in Farriol et al. 2007; details of other interventions and social agreements in Otero 2011a; see also Otero 2010: 153–156).

The development of this land management scheme was accompanied by the maintenance of strategic forest tracks and water infrastructures (to facilitate the work of fire engines) as well as by patrolling shifts by *ADF*'s team of volunteers in pick-ups equipped with water. These activities were part of official municipal and regional wildfire prevention schemes and were coordinated with the Catalan Fire Department which is the ultimate responsible for wildfire suppression. *ADF*'s land management scheme was also accompanied by outreach activities to raise public awareness on wildfire risk and prevention measures, notably in collaboration with the association Local Environmental Council. All these efforts were likewise integrated with a research project aimed at activating the social memory of the

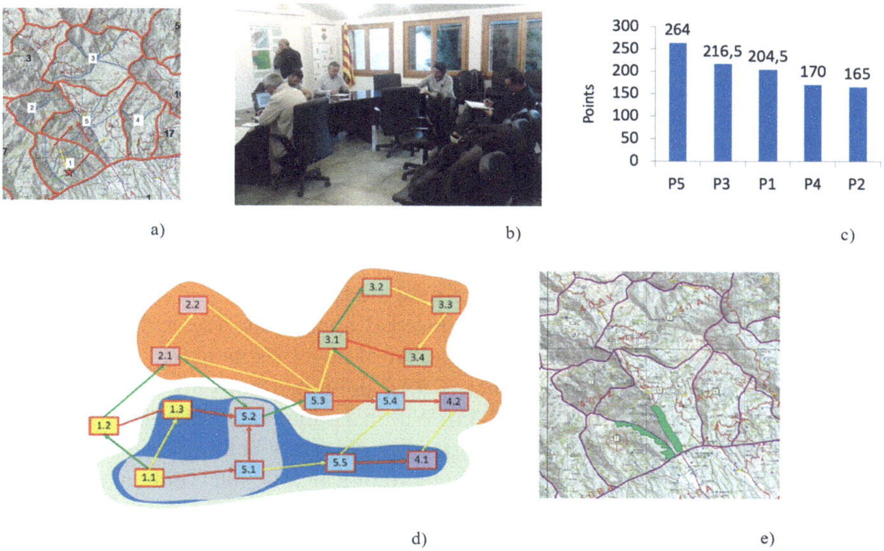

Fig. 7.3 Case study #2: Participatory wildfire planning network in the region Montseny-Montnegre-Corredor. (**a**) The pilot area was divided in wildfire containment polygons (red line) and sub-polygons (blue line). A convective wildfire was simulated starting in polygon 1 (red star) and driven by west winds. Red arrow: front spread (high intensity); yellow arrow: flank spread (medium intensity); green arrow: back spread (low intensity). (**b**) Public exhibition in Montseny municipality (4 May 2016) co-organized by the local government. After reading posters about the participatory process and the values of each polygon, participants were asked to fill a form where they could rank the polygons and add additional values. The posters were previously co-designed with regional actors. (**c**) Aggregate polygon ranking, considering the two exhibitions conducted. Overall, 68 persons voted. (**d**) Potential area burnt according to different intervention scenarios. Grey, blue and green areas correspond to a wildfire starting in sub-polygon 1.1 (discussed with the actors). (**e**) Strategic Management Areas (SMA) necessary to protect the most valued polygons. The SMA in polygon 5 would allow to contain the wildfire in the valley bottom. If this would work, the burnt area would resemble the grey shape in (**d**). If this SMA is not developed, the burnt area would resemble the blue or green shapes in (**d**). The SMA between polygons 1 and 5 would help by reducing wildfire spotting distance (see details in Otero et al. 2018). (Source: Otero et al. 2018 (**a**, **c**, **d**, **e**); I. Otero (**b**))

municipality's rural past, conducted with the Municipal Archives (Aguilar 2012; Ruiz et al. 2008). Such a research project was part of the PhD thesis of the author (Otero et al. 2011) at the Institute of Environmental Science and Technology of the Autonomous University of Barcelona (ICTA-UAB).

Case study #2 is located in the north-east of the BMR, in a region spanning from the southern slopes of the pre-coastal Montseny mountain range to the coastal range of Montnegre-Corredor, including municipalities of the counties of Vallès Oriental and Maresme (Table 7.1). This region is a hotspot of industrial activities, strategic transport infrastructures and urban and suburban areas, including touristic towns by the seaside. The author moved to this region by the end of 2012. In 2014 he started a postdoc on the political ecology of wildfires in Humboldt University of Berlin, which allowed him to set up and monitor a participatory network to build resilience to wildfires in the mentioned region. For this, he built on the experience he had acquired in Matadepera but also integrated the expertise of two key persons: the head of *GRAF* and a social activist from Barcelona. The former aimed at democratizing the practices of the Fire Department by incorporating social values of landscape into wildfire management. The latter aimed at empowering citizens in democratic urban and land-planning politics. It was the aftermath of Spain's acute economic and democratic crisis, when citizens mobilized in new practices of *real democracy* and in defence of public services (Asara and Muraca 2015).

The method shared the philosophy of case study #1 developed in Matadepera: wildfire intensity and impacts can be reduced if appropriate fuel reduction treatments are implemented in SMA. Yet, a number of novelties increased its participatory potential and its complexity. First, *GRAF* delimited not only SMA but also wildfire containment polygons (Fig. 7.3a). Within a polygon, wildfire is expected to burn with a similar behaviour, and all together polygons help assess the wildfire potential size according to different meteorological and intervention scenarios (Fig. 7.3d; Castellnou et al. 2019). SMA and polygons were drawn with the specific scenario of a convective wildfire driven by west winds in mind. Second, a method was co-designed with regional actors (including wildfire and forest management agencies, associations of forest owners and natural parks) to rank these polygons by citizens according to a diversity of landscape values (Fig. 7.3b). Third, the selection of which SMA would be implemented was done according to the ranking stemming from the participatory process (Fig. 7.3c). SMA protecting the most valued polygons were thus prioritized (Fig. 7.3e). And finally, the method included deliberation sessions whereby the Fire Department could negotiate its wildfire management strategy with those actors and citizens having a stake in it. For example, in the last public session of the project, the two prioritized SMA were agreed upon as part of a strategy to limit wildfire spread in the most valued polygon (Fig. 7.3d, e). Due to time and resource constraints, this participatory process was only implemented in a pilot area within the region, consisting of four municipalities in Montseny mountain range (see details of the process in Otero et al. 2018).

The resulting participatory network was mostly active between 2014 and 2016. During this period, the author was partly living in the region (to conduct participatory activities and fieldwork) and partly in Berlin. In 2017 and 2018, the activities of the network were limited to dissemination purposes and to the preparation of a project for an extended phase. In 2018 the author moved to Switzerland and started working in the University of Lausanne.

The extended phase of case study #2 is case study #3, which was funded by the call for climate change adaptation projects of the *Fundación Biodiversidad* (Spanish Ministry for Ecological Transition) and hosted by ICTA-UAB. It focused on the Montseny Biosphere Reserve, in the north-eastern fringe of the BMR. Out of the Reserve's 18 municipalities, 11 are located in the BMR (county of Vallès Oriental), and 7 are located outside (counties of Selva and Osona). It has a population of 51,573 people and is partly protected by the Montseny Natural Park (Table 7.1). This project aimed at improving the pilot method developed in case study #2. For

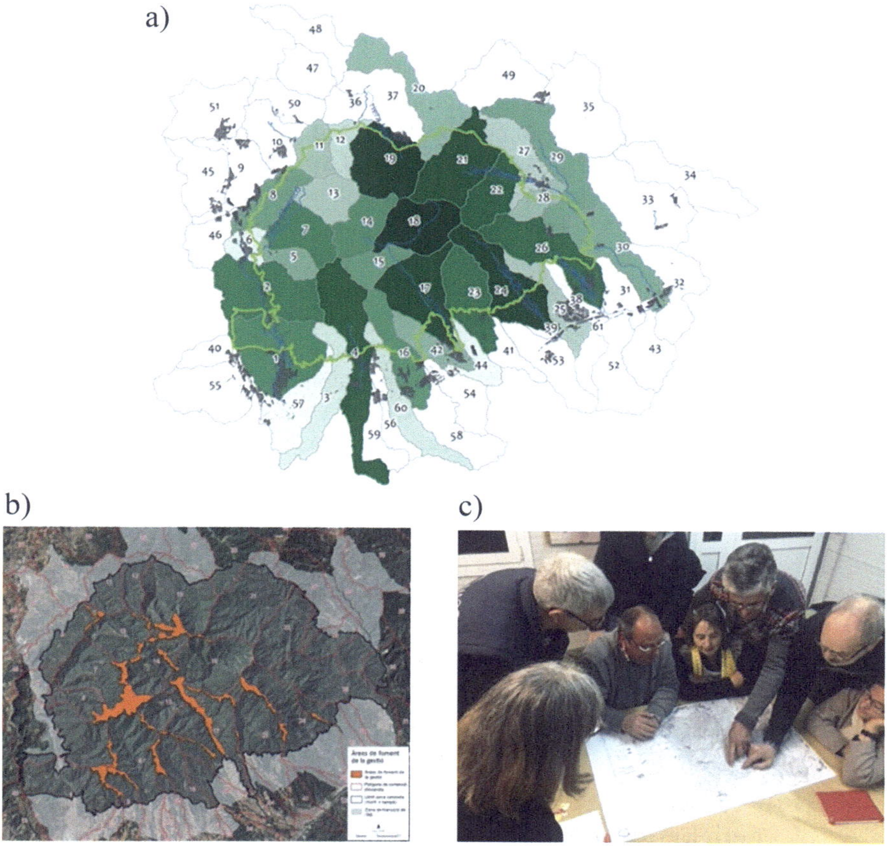

Fig. 7.4 Case study #3: participatory wildfire planning network in Montseny Biosphere Reserve (MBR). (**a**) Multi-criteria prioritization of wildfire containment polygons considering all value categories (biodiversity, socioeconomic activities, heritage, leisure, other values and vulnerability to climate change). Darker green indicates higher priority. The bright green line indicates the perimeter of the valued region (corresponding to the core and buffer areas of MBR). (**b**) Strategic Management Areas (orange) to reduce the impacts of a west wind-driven convective wildfire in the most valued polygons. Red lines: polygons. Dark grey: valued region (core and buffer areas of MBR). Light grey: non-valued region (transition area of MBR). (**c**) Meeting with local actors in Arbúcies (6 February 2019). Two meetings like this one were held with Groups of Forest Defence (*Agrupacions de Defensa Forestal, ADF*), firefighters and landowners as well as economic actors related to agriculture and tourism, in order to identify high-value areas for the category *Other values*. (Source: (Otero et al. 2020b (**a**, **b**)); *Espurna* project (**c**))

this, the author hired a post-doctoral researcher at ICTA-UAB and supervised the project from Switzerland. Project partners included the *Pau Costa Foundation*, a spin-off of *GRAF* aimed at disseminating expert knowledge on fire ecology and management, and *El Risell*, a cooperative dealing with participatory processes in urban and land management.

As in case study #2, case study #3 included a wide diversity of actors such as public agencies in charge of wildfire prevention and suppression, natural parks, town councils, an association of forest landowners, *ADF*, museums etc. (Fig. 7.4c). Regarding the input of wildfire experts, it used the same approach as in case study #2 (delimitation of wildfire containment polygons and SMA specific to convective wildfires driven by west winds), but it also included a detailed analysis of the wildfire types that affected the region in the past, as well as a qualitative assessment of the expected trends with climate change. Regarding the participatory landscape valuation, the main difference with case study #2 was the creation of raster maps that assigned a score to each pixel per different value categories. A multi-criteria evaluation method was then used to rank polygons according to the quantity and diversity of values contained (Fig. 7.4a). As in case study #2, the wildfire experts of the team planned a wildfire prevention and suppression strategy intended to protect the priority polygons, this time at the scale of the whole Montseny Biosphere Reserve (Fig. 7.4b; see details of the process in Otero et al. 2020b; Gamboa et al. under review). This network implemented most of its activities between September 2018 and June 2019, coinciding with the funding period. Afterwards, activities focused on disseminating the project results as well as on trying to integrate them into the wildfire prevention planning of the Montseny Natural Park. The latter hired the post-doctoral researcher co-leading the network to work on a diagnosis of its livestock sector, which is considered key for both wildfire prevention and biodiversity conservation. However, the SMA planned in both case studies #2 and #3 have not yet been implemented.

7.5 How to Enhance the Transformative Capacity of Participatory Wildfire Governance?

This section explores how can we enhance the transformative capacity of participatory wildfire governance. It does so by analysing each case study with regard to (a) its achievements, (b) the challenges faced and how they were dealt with and (c) possible new transformative avenues emerging from the experiences.

In Matadepera (case study #1), a bottom-up land management scheme to build resilience to wildfires has been ongoing for 18 years. The scheme has been designed and implemented by a local association led by volunteers which de facto assumes the competence of wildfire prevention that by law belongs to the Town Council. This suggests that a locally rooted and empowered civil society can have a key role in the development of official long-term resilience policies, especially when a large disturbance (2003 wildfire) triggers social organization and learning to adapt (Otero 2010; Folke et al. 2005). This supports the findings from other studies on the importance of people-place connections and community cohesion for resilience (Prior and Eriksen 2013; Berkes and Ross 2013; McCaffrey 2015). Of course, a community is hardly fully cohesive. Matadepera is a wealthy residential area of the BMR, with luxury houses spreading in the hillside of Sant Llorenç del Munt-l'Obac mountain range. Yet there are large differences within the local population, notably between the old town centre and the residential developments of the WUI. By building resilience to wildfires on a voluntary basis, the local Group of Forest Defence (*ADF*) helps maintain the economic value of the elite WUI which lies at the heart of the municipality's vulnerability. Thus, the author asked himself in several occasions: *Who are we [ADF] working for?* But he concluded that working for the protection of the local landscape and people is a worthwhile endeavour per se, regardless of questions of environmental justice.

In addition, continuous outreach on the management scheme and the activation of the rural memory of the town through research and dissemination (Aguilar 2012) allowed to spread the understanding among the local population that a high wildfire risk stems from the urbanization of the countryside (and related afforestation) and that it requires a holistic land management rather than only suppression means. Among those in *ADF* leading the scheme, such reframing was accompanied by a philosophy consisting in *not to exclude wildfire from the local ecosystem, but rather to build a social-ecological system able to absorb fire events and retain essential structures, processes, and feedbacks* (Otero 2011a: 3). Thus, case study #1 illustrates how the resilience approach to wildfires and fire restoration can be implemented in a territory. In particular, it allows understanding that such implementation is not automatic but requires a negotiation with a diversity of actors who can facilitate or oppose the new approach. In the first years of the scheme's implementation in Matadepera, the natural park reluctantly authorized the forestry works because it feared being criticized for excessive forest clearance by the park's visitors and the inhabitants of the WUI. The absence of strong criticism and the healthy development of the thinned forests in the following years made the natural park more supportive of these interventions. However, some citizens still consider that those works had unacceptable impacts on pine woodlands. In 2013, *ADF* and the wildfire specialist group of the Catalan Fire Department (*GRAF*) planned a prescribed burning within the borders of the natural park, but the latter did not give a favourable report arguing that the park's management scheme did not allow for this particular use of fire and that the smoke would trouble the inhabitants of the WUI. Thus, the burning had to be aborted. As of 2021, *ADF* and *GRAF* are planning another burning outside the borders of the natural park in order to showcase the benefits of this wildfire prevention practice without depending on the natural park's favourable report. These interactions with actors reveal the need to better consider trade-offs between different ecosystem services and values when planning wildfire resilience measures, as well as to integrate antagonist actors earlier on (Otero et al. 2018; Depietri and Orenstein 2020). The methods of case studies #2 and #3 were precisely developed to integrate a broader set of ecosystem services and values, from biodiversity conservation to social recreation, as will be discussed later.

Another significant challenge found in Matadepera was related to the management of the flock. Public money was invested to support the shepherd so that he could graze in the thinned forests and help control shrub growth. Yet the shepherd prioritized other grazing areas that suited better the flock's nutritional needs. In addition, animals escaped relatively often, creating conflicts with the forest owners and the inhabitants of the WUI. *ADF* thus faced pressure from public authorities and citizens and had to invest considerable efforts to appease tensions while justifying the choice of extensive grazing for wildfire resilience. After several consecutive agreements with different shepherds that did not work, around 2013 an agreement was signed with a shepherd which finally allowed to balance the flock's needs with those related to wildfire management. Currently, this shepherd still grazes his flock in Matadepera and sells the meat in the region. This experience illustrates that the recovery of the traditional rural mosaic in order to better coexist with wildfire (Loepfe et al. 2010, 2012; Aquilué et al. 2020) is not an automatic endeavour. Rather, in highly suburbanized contexts like Matadepera, such a recovery is likely to require a long trial-and-error process to weave new institutional, social and social-ecological relationships.

Knowing whether the management scheme implemented by *ADF* actually contributed to resilience was another major challenge. *ADF* only managed to implement a 2-year monitoring scheme on the grazing effects in the understorey, indicating an incipient change of species composition and a certain (low) control on plant growth (Bartolomé 2009). However, the absence of a long-term monitoring programme on the overall effects of the scheme on wildfire potential intensity and severity made it impossible to made positivist claims on its effectiveness. It is however likely that the current vulnerability is lower than a hypothetical no-intervention scenario, especially considering recent climate warming. This points at the need to further develop methods and languages to assess the (transformative) effects of participatory wildfire governance, a need that has been also identified for transdisciplinary research programmes (Schäfer et al. 2020; Otero et al. 2020a).

The planning network implemented in case study #2 was a significant step forward regarding case study #1 for at least two reasons. First, a broad set of values from actors and citizens (biodiversity, cultural heritage, socioeconomic activities etc.) was integrated in the design of measures to build resilience, potentially appeasing trade-offs between ecosystem services and values perceived by different actors (Depietri and Orenstein 2020). And second, the interaction between citizens, actors and the Fire Department was planned in such a way that the latter could use the results in their suppression activities. This means, for example, that if there is a wildfire in the area, *GRAF* could focus its efforts in protecting the most valued containment polygons (Otero et al. 2018). Wildfire containment polygons are territorial units used by the Fire Department to assess wildfire potential size and impacts, within which wildfire is expected to burn with a similar behaviour (Castellnou et al. 2019; Fig 7.3a). The integration of the social values and interests at stake into operational wildfire strategies is an emerging need among Fire Departments as they find it increasingly difficult to make decisions about intervention priorities, especially under extreme conditions (Castellnou 2017). In this sense, the prioritization of wildfire containment polygons based on an aggregate score (as done in case study #2) is a simple way to translate complex values into wildfire management that can complement existing planning approaches (Everett 2002; Tàbara et al. 2003; Morehouse et al. 2010).

These achievements were enabled by the co-design principle, based on reaching agreements with actors on what was to be done and how during the entire process. This implied that actors were allowed to shape the process according to their own interests, legitimizing it along the way (Otero et al. 2018). Remarkably, a general consensus on priority polygons was found among heterogeneous sets of actors and citizens, suggesting that the conflicting nature of wildfire policies reported by some studies (Carroll et al. 2005, 2006; González-Hidalgo et al. 2014) might be appeased by means of integrative planning networks. In addition, citizens reported to have acquired some knowledge on wildfire and wildfire management after participating in the exhibitions. This suggests that the method developed in case study #2 could be a way to recover local fire knowledge in metropolitan contexts, a crucial component of resilience that nevertheless vanished with urbanization (Huffman 2013). Legitimacy and learning seem to be two crucial processes that helped opening up *GRAF*'s expert knowledge (and its potential effects on landscape) to social deliberation, and they should be properly considered in new transformative wildfire governance efforts.

However, challenges were also huge. It was not possible to integrate all actors and values involved in or potentially affected by wildfire. This was due to not only the project's limited resources but also the extreme magnitude of the expected wildfires, which could for instance block the highway that crosses the region, causing significant socioeconomic impacts elsewhere. Moreover, in the network's final session with actors, *GRAF* did not commit to fully protect the most valued polygon, arguing that the wildfire intensity would be beyond their workable threshold in half of it (Fig. 7.3a, red arrows). In other words, it is not only the local community and the regional actors that have a voice in shaping the landscape of the future but also wildfire itself. In that session though, this insight did not raise opposition by participants, suggesting that some actors in metropolitan settings may be ready to accept letting some wildfires burn. Actually, the method developed in case study #2 could be used to operationalize fire restoration calls (Donovan and Brown 2007; Regos et al. 2014; North et al. 2015; Thompson et al. 2015; Domènech et al. 2018; Oliveres 2021) by agreeing upon which wildfires will be left to burn, under what circumstances and according to whose values between the Fire Department and the affected communities. Other actors that could not be successfully integrated in case study #2 include a network of alternative land management projects in the region. Such network was started by the author and his commune's housemates in order to enhance the transformative potential of case study #2. These projects were considered to have the potential to shape a less flammable landscape and to reframe the wildfire issue as a political problem requiring a transformation of the current land-use model, characterized by persisting land abandonment and urbanization. Projects included for instance producers of artisan goat cheese, organic meat (sheep/goat) and firewood. This network was expected to converge with the landscape co-valuation, but it was aborted due to resource constraints. This suggests that the integration of actors should be better designed in advance and the necessary resources allocated accordingly.

The method developed in case study #3 was a refined version of case #2. Whereas polygons in #2 were ranked based on value scores that were not spatially explicit, in #3 the valuation was done with a GIS (raster) where each pixel had a value score. These value maps can be overlapped with different wildfire containment polygon maps, i.e. designed for different wildfire types and climate change scenarios (Otero et al. 2020b). In turn, the participatory multi-criteria evaluation of polygons can be updated in successive phases of public deliberation, either by adding value categories or by adjusting their weights (Gamboa et al. under review). This can help adapt wildfire management to both changing risk and social priorities, complementing other planning approaches that integrate the biophysical and social dimensions of risk (Ager et al. 2015; Fischer et al. 2016).

Case study #3 achieved a considerable social participation in terms of diversity of public and private actors (Otero et al. 2020b). But municipal elections occurred in May 2019. This complicated the involvement of the mayors in the organization of exhibitions for the participation of local communities, something that instead worked very well in case study #2 (Fig. 7.3b, c; Otero et al. 2018). In both cases though, the selected SMA could not be integrated in the wildfire planning schemes of the municipalities and the natural park, and they have not (yet) been implemented. The reason for this is probably that the network started by the author did not take sufficient root in the region, where unlike case study #1, he was an outsider scientist living elsewhere. This highlights how precious endogenous social initiatives (like case study #1) are for community resilience to wildfires. These initiatives connect expert and motivated volunteers with decision-makers (e.g. the Town Council) which can integrate the measures developed by the former into official land management schemes endowed with sufficient financial resources. The experience in #2 and #3 shows that new transformative wildfire planning projects should plan an executive phase to fund the implementation of those SMA selected in the valuation phase. This could be done in collaboration with local governments and the natural park, something that would require a better transdisciplinary co-design of the project (Lang et al. 2012; Moser 2016). Moreover, the discussions with actors in case study #3 highlighted that one of the main obstacles for the implementation of SMA was the low profitability of the primary sector (agriculture, ranching and forestry). This needs to be addressed when co-designing new transformative projects, for example, by including payments for ecosystem services or subsidies for farmers, shepherds and forest owners that provide fire-regulating services via the implementation of SMA (Varela et al. 2018; Depietri and Orenstein 2019).

Finally, case study #3 revealed one crucial challenge for transformative wildfire governance. Towards the end of the project, the team realized that information on social values and priority polygons was not enough for the Fire Department to decide how to manage a wildfire. In extreme wildfires that are beyond suppression capacity, protecting one polygon amounts to losing others. Even if the GIS would allow to calculate the values gained and lost under different management decisions, future projects should also ask participants what would they be willing to lose in order to protect their most valued areas (Otero et al. 2020b). This seems especially relevant if the proposal to use wildfires as a fuel management strategy (Regos et al.

2014; Domènech et al. 2018; Oliveres 2021) is to achieve social acceptance, and it will need to be tested in a new phase of action research.

7.6 Conclusions

A new type of extreme wildfires, with unprecedented intensity and speed, is being observed globally. Chile 2017, Portugal 2017, Sweden 2018 and Australia 2020 are just some examples of the magnitude of these events which produce casualties and have high socioeconomic and ecological impacts. These new events come to exacerbate the vulnerability of many wildfire-prone regions, especially those with extensive cities and metropolitan areas in the WUI. Fire departments, civil protection agencies and researchers are therefore increasingly concerned. Some of them have long stressed the limits of wildfire suppression, arguing that this policy enables fuel build-up and increases wildfire intensity. They have instead pointed at the need to adapt to this perturbation by creating resilient social-ecological systems that are able to coexist with wildfire. Seen from a holistic point of view, the changes suggested by the literature to coexist with wildfire hint at a profound transformation of the current social-ecological configurations.

This chapter reflects on 18 years of participatory wildfire governance in the BMR (Spain) with a transformative lens. The analysis of three projects where the author was involved during these years shed light on how the changes suggested by the literature play out in a concrete geographical setting, as well as some of the challenges ahead. In particular, the chapter shows the crucial role that a locally rooted civil society can have when it is able to network with key agencies and actors over the long term. Indeed, a long-term commitment well beyond electoral cycles is necessary to weave new social, institutional and social-ecological relationships conducive to resilient systems. The analysis also shows the importance of developing integrative planning networks where different ecosystem services and values are considered in successive phases of public deliberation. This can enable a dialogue between the fire departments, agencies, local governments and communities that facilitates wildfire restoration, even if some challenges remain regarding the integration of social priorities into operational decisions. Finally, deepening the transdisciplinary content of participatory wildfire governance can increase its transformative potential.

Acknowledgements There is a large number of persons and institutions to whom I am grateful for their strong and persistent support throughout these 18 years. The publications referred to in Table 7.1 provide their names and affiliations. My special thanks go to Ricard Farriol, Toti Garcia, Joan Trenchs, Marc Castellnou, Etel Arilla, Itziar González, Jonas Ø. Nielsen and Gonzalo Gamboa. I am also grateful to the editors of this book for their feedback on an earlier version of the chapter.

References

Abrams JB, Knapp M, Paveglio TB et al (2015) Re-envisioning community-wildfire relations in the U.S. West as adaptive governance. Ecol Soc 20(34). https://doi.org/10.5751/ES-07848-200334

ADF Matadepera (Agrupació de Defensa Forestal de Matadepera) (2005) Projecte d'actuacions per a la prevenció de grans incendis forestals al municipi de Matadepera. Unpublished project

Ager AA, Kline JD, Fischer AP (2015) Coupling the biophysical and social dimensions of wildfire risk to improve wildfire mitigation planning. Risk Anal 35:1393–1406. https://doi.org/10.1111/risa.12373

Aguilar R (2012) Memòries d'una feixa. La urbanització d'un poble pagès de la regió de Barcelona (Matadepera, 1931–1983). Ecos Sccl, Barcelona

Albertson K, Aylen J, Cavan G et al (2010) Climate change and the future occurrence of moorland wildfires in the Peak District of the UK. Clim Res 45:105–118. https://doi.org/10.3354/cr00926

Almstedt Å, Reed MG (2013) Introducing a framework for good and adaptive governance: an application to fire management planning in Canada's boreal forest. For Chron 89:664–674. https://doi.org/10.5558/tfc2013-119

Aquilué N, Fortin M-J, Messier C et al (2020) The potential of agricultural conversion to shape forest fire regimes in Mediterranean landscapes. Ecosystems 23:34–51. https://doi.org/10.1007/s10021-019-00385-7

Asara V, Muraca B (2015) Indignados (Occupy!). In: D'Alisa G, Demaria F, Kallis G (eds) Degrowth. A vocabulary for a new era. Routledge, pp 169–171

Badia A, Pallares-Barbera M, Valldeperas N et al (2019) Wildfires in the wildland-urban interface in Catalonia: vulnerability analysis based on land use and land cover change. Sci Total Environ 673:184–196. https://doi.org/10.1016/j.scitotenv.2019.04.012

Bartolomé J (2009) Efectes de la pastura amb bestiar oví sobre les espècies llenyoses de sotabosc en les franges de protecció de grans incendis forestals de Matadepera. Estructura de la vegetació. Unpublished report

Başnou C, Álvarez E, Bagaria G et al (2013) Spatial patterns of land use changes across a Mediterranean metropolitan landscape: implications for biodiversity management. Environ Manag 52:971–980. https://doi.org/10.1007/s00267-013-0150-5

Berkes F, Ross H (2013) Community resilience: toward an integrated approach. Soc Nat Resour 26:5–20. https://doi.org/10.1080/08941920.2012.736605

Boada M, Zahonero A (1998) Medi ambient, una crisi civilitzadora. La Magrana, Barcelona

Bowman DMJS, Williamson GJ, Abatzoglou JT et al (2017) Human exposure and sensitivity to globally extreme wildfire events. Nat Ecol Evol 1:1–6. https://doi.org/10.1038/s41559-016-0058

Brotons L, Aquilué N, de Cáceres M et al (2013) How fire history, fire suppression practices and climate change affect wildfire regimes in Mediterranean landscapes. PLoS One 8:e62392. https://doi.org/10.1371/journal.pone.0062392

Brotons L, Duane A (2019) Correspondence: uncertainty in climate-vegetation feedbacks on fire regimes challenges reliable long-term projections of burnt area from correlative models. Fire 2:8. https://doi.org/10.3390/fire2010008

Buizer M, Kurz T (2016) Too hot to handle: Depoliticisation and the discourse of ecological modernisation in fire management debates. Geoforum 68:48–56. https://doi.org/10.1016/j.geoforum.2015.11.011

Butler WH, Goldstein BE (2010) The US Fire Learning Network: springing a rigidity trap through multiscalar collaborative networks. Ecol Soc 15:21

Calkin DE, Thompson MP, Finney MA (2015) Negative consequences of positive feedbacks in US wildfire management. For Ecosyst 2:9. https://doi.org/10.1186/s40663-015-0033-8

Campos JC, Bernhardt J, Aquilué N et al (2021) Using fire to enhance rewilding when agricultural policies fail. Sci Total Environ 755:142897. https://doi.org/10.1016/j.scitotenv.2020.142897

Carroll MS, Cohn PJ, Seesholtz DN et al (2005) Fire as a galvanizing and fragmenting influence on communities: the case of the Rodeo–Chediski fire. Soc Nat Resour 18:301–320. https://doi.org/10.1080/08941920590915224

Carroll MS, Higgins LL, Cohn PJ et al (2006) Community wildfire events as a source of social conflict. Rural Sociol 71:261–280. https://doi.org/10.1526/003601106777789701
Cash D, Adger WN, Berkes F et al (2006) Scale and cross-scale dynamics: governance and information in a multilevel world. Ecol Soc 11:8. https://doi.org/10.5751/ES-01759-110208
Castellnou M (2020) Extreme wildfire events under climate change: what to expect on Central/Northern Europe. PyroLife International Symposium, 3 June 2020. Available at https://www.youtube.com/watch?v=IbE_F32LHmk
Castellnou M (2017) Talk at the international congress on prescribed fires. Barcelona, 1–3 February 2017. Available at https://www.youtube.com/watch?v=vrnHo6LAj8o (see Q&A)
Castellnou M, Prat-Guitart N, Arilla E et al (2019) Empowering strategic decision-making for wildfire management: avoiding the fear trap and creating a resilient landscape. Fire Ecol 15:31. https://doi.org/10.1186/s42408-019-0048-6
Catalán B, Saurí D, Serra P (2008) Urban sprawl in the Mediterranean? Patterns of growth and change in the Barcelona Metropolitan Region 1993–2000. Landsc Urban Plan 85:174–184. https://doi.org/10.1016/j.landurbplan.2007.11.004
Chapin FS, Trainor SF, Huntington O et al (2008) Increasing wildfire in Alaska's boreal forest: pathways to potential solutions of a wicked problem. Bioscience 58:531–540. https://doi.org/10.1641/B580609
Darbellay F, Moody Z, Sedooka A et al (2014) Interdisciplinary research boosted by serendipity. Creat Res J 26:1–10. https://doi.org/10.1080/10400419.2014.873653
Depietri Y, Orenstein DE (2020) Managing fire risk at the wildland-urban interface requires reconciliation of tradeoffs between regulating and cultural ecosystem services. Ecosyst Serv 44:101108. https://doi.org/10.1016/j.ecoser.2020.101108
Depietri Y, Orenstein DE (2019) Fire-regulating services and disservices with an application to the Haifa-Carmel Region in Israel. Front Environ Sci 7:107. https://doi.org/10.3389/fenvs.2019.00107
Doerr SH, Santín C (2016) Global trends in wildfire and its impacts: perceptions versus realities in a changing world. Phil Trans R Soc B 371:20150345. https://doi.org/10.1098/rstb.2015.0345
Domènech R, Piqué M, Beltrán M et al (2018) El paper del foc en la conservació de l'hàbitat dels boscos de pinassa (*Pinus nigra* Arn.) Project report LIFE+ PINASSA. Centre de Ciència i Tecnologia Forestal de Catalunya and Bombers de la Generalitat de Catalunya. Available at http://lifepinassa.eu/wp-content/uploads/2018/06/Manual_fuego_CAT.pdf
Donovan GH, Brown TC (2007) Be careful what you wish for: the legacy of Smokey Bear. Front Ecol Environ 5:73–79. https://doi.org/10.1890/1540-9295(2007)5[73:BCWYWF]2.0.CO;2
Duane A, Aquilué N, Canelles Q et al (2019) Adapting prescribed burns to future climate change in Mediterranean landscapes. Sci Total Environ 677:68–83. https://doi.org/10.1016/j.scitotenv.2019.04.348
Duane A, Brotons L (2018) Synoptic weather conditions and changing fire regimes in a Mediterranean environment. Agric For Meteorol 253–254:190–202. https://doi.org/10.1016/j.agrformet.2018.02.014
Duane A, Castellnou M, Brotons L (2021) Towards a comprehensive look at global drivers of novel extreme wildfire events. Clim Chang 165:43. https://doi.org/10.1007/s10584-021-03066-4
Eriksen C (2014) Gendered risk engagement: challenging the embedded vulnerability, social norms and power relations in conventional Australian bushfire education. Geogr Res 52:23–33. https://doi.org/10.1111/1745-5871.12046
Everett Y (2002) Community participation in fire management planning: a case from California, USA. In: Moore P, Ganz D, Tan LC et al (eds) Communities in flames: proceedings of an international conference on community involvement in fire management. FAO and FireFight South East Asia, Bangkok, pp 125–133
Everett Y, Fuller M (2011) Fire Safe Councils in the interface. Soc Nat Resour 24:319–333. https://doi.org/10.1080/08941920903313835
Farriol R, Plana E, Otero I (2007) Integració de la gestió del territori i la planificació forestal a la gestió del risc d'incendis forestals. El cas del Projecte d'actuacions per a la prevenció de grans

incendis forestals al municipi de Matadepera (Vallès Occidental). In: VI Trobada d'Estudiosos de Sant Llorenç del Munt i l'Obac. Diputació de Barcelona, Barcelona, pp 17–23
Fischer A, Jasny L (2017) Capacity to adapt to environmental change: evidence from a network of organizations concerned with increasing wildfire risk. Ecol Soc 22:23. https://doi.org/10.5751/ES-08867-220123
Fischer AP, Spies TA, Steelman TA et al (2016) Wildfire risk as a socioecological pathology. Front Ecol Environ 14:276–284. https://doi.org/10.1002/fee.1283
Flannigan M, Cantin AS, de Groot WJ et al (2013) Global wildland fire season severity in the 21st century. For Ecol Manag 294:54–61. https://doi.org/10.1016/j.foreco.2012.10.022
Folke C, Hahn T, Olsson P et al (2005) Adaptive governance of social-ecological systems. Annu Rev Environ Resour 30:441–473. https://doi.org/10.1146/annurev.energy.30.050504.144511
Gamboa G, Otero I, Bueno C et al (under review) Participatory multicriteria evaluation of landscape values to inform wildfiremanagement. Where is the added value? J Environ Manage
Gazzard R, McMorrow J, Aylen J (2016) Wildfire policy and management in England: an evolving response from Fire and Rescue Services, forestry and cross-sector groups. Phil Trans R Soc B 371:20150341. https://doi.org/10.1098/rstb.2015.0341
González-Hidalgo M, Otero I, Kallis G (2014) Seeing beyond the smoke: the political ecology of fire in Horta de Sant Joan (Catalonia). Environ Plan A 46:1014–1031. https://doi.org/10.1068/a45600
González-Hidalgo M, Otero I, Kallis G (2013) Más allá del humo: La ecología política de los incendios forestales a partir del caso de Horta de Sant Joan (Tarragona, Cataluña). Doc Anàlisi Geogràfica 59:21–50
Herr K, Anderson G (2005) The action research dissertation. A guide for students and faculty. SAGE Publications
Howitt R (2014) Coexisting with fire? A commentary on the scale politics of adaptation. Geogr Res 52:61–64. https://doi.org/10.1111/1745-5871.12036
Huffman MR (2013) The many elements of Traditional Fire Knowledge: synthesis, classification, and aids to cross-cultural problem solving in fire-dependent systems around the world. Ecol Soc 18:3. https://doi.org/10.5751/ES-05843-180403
IERMB (2021) La metròpoli en 100 indicadors. L'AMB en xifres 2020. Available at https://iermb.uab.cat/ca/amb-en-xifres/la-metropoli-en-100-indicadors-lamb-en-xifres-2020/. Accessed 11 Aug 2021
Jolly WM, Cochrane MA, Freeborn PH et al (2015) Climate-induced variations in global wildfire danger from 1979 to 2013. Nat Commun 6:7537. https://doi.org/10.1038/ncomms8537
Kelly LT, Brotons L (2017) Using fire to promote biodiversity. Science 355:1264–1265. https://doi.org/10.1126/science.aam7672
Kelly LT, Giljohann KM, Duane A et al (2020) Fire and biodiversity in the Anthropocene. Science 370. https://doi.org/10.1126/science.abb0355
Krawchuk MA, Moritz MA, Parisien M-A et al (2009) Global pyrogeography: the current and future distribution of wildfire. PLoS One 4:e5102. https://doi.org/10.1371/journal.pone.0005102
Lang DJ, Wiek A, Bergmann M et al (2012) Transdisciplinary research in sustainability science: practice, principles, and challenges. Sustain Sci 7:25–43. https://doi.org/10.1007/s11625-011-0149-x
Loepfe L, Martinez-Vilalta J, Oliveres J et al (2010) Feedbacks between fuel reduction and landscape homogenisation determine fire regimes in three Mediterranean areas. For Ecol Manag 259:2366–2374. https://doi.org/10.1016/j.foreco.2010.03.009
Loepfe L, Martinez-Vilalta J, Piñol J (2012) Management alternatives to offset climate change effects on Mediterranean fire regimes in NE Spain. Clim Chang 115:693–707. https://doi.org/10.1007/s10584-012-0488-3
Marull J, Pino J, Mallarach JM et al (2007) A land suitability index for strategic environmental assessment in metropolitan areas. Landsc Urban Plan 81:200–212. https://doi.org/10.1016/j.landurbplan.2006.11.005

Marull J, Pino J, Tello E et al (2010) Social metabolism, landscape change and land-use planning in the Barcelona Metropolitan Region. Land Use Policy 27:497–510. https://doi.org/10.1016/j.landusepol.2009.07.004

McCaffrey S (2015) Community wildfire preparedness: a global state-of-the-knowledge summary of social science research. Curr For Rep 1:81–90. https://doi.org/10.1007/s40725-015-0015-7

McKenzie D, Gedalof Z, Peterson DL et al (2004) Climatic change, wildfire, and conservation. Conserv Biol 18:890–902. https://doi.org/10.1111/j.1523-1739.2004.00492.x

Moore P, Ganz D, Tan LC et al (eds) (2002) Communities in flames: proceedings of an international conference on community involvement in fire management. FAO and FireFight South East Asia, Bangkok

Morehouse BJ, O'Brien S (2008) Facilitating public involvement in strategic planning for wildland fire management. Prof Geogr 60:495–507. https://doi.org/10.1080/00330120802211836

Morehouse BJ, O'Brien S, Christopherson G et al (2010) Integrating values and risk perceptions into a decision support system. Int J Wildland Fire 19:123–136. https://doi.org/10.1071/WF08064

Moritz MA, Batllori E, Bradstock RA et al (2014) Learning to coexist with wildfire. Nature 515:58–66. https://doi.org/10.1038/nature13946

Moritz MA, Parisien M-A, Batllori E et al (2012) Climate change and disruptions to global fire activity. Ecosphere 3:1–22. https://doi.org/10.1890/ES11-00345.1

Moser SC (2016) Can science on transformation transform science? Lessons from co-design. Curr Opin Environ Sustain 20:106–115. https://doi.org/10.1016/j.cosust.2016.10.007

Müller MM, Vilà-Vilardell L, Vacik H et al (2020) Forest fires in the Alps. State of knowledge, future challenges and options for an integrated fire management (White paper for policy makers). EUSALP Action Group 8. Available at https://www.dora.lib4ri.ch/wsl/islandora/object/wsl%3A22589/

North MP, Stephens SL, Collins BM et al (2015) Reform forest fire management. Science 349:1280–1281. https://doi.org/10.1126/science.aab2356

Oliveres J (2021) The management of fire regime complexity at the edge of criticality and global change (PhD Thesis). Universitat Politècnica de Catalunya, Spain

Olsson P, Folke C, Galaz V et al (2007) Enhancing the fit through adaptive co-management: creating and maintaining bridging functions for matching scales in the Kristianstads Vattenrike Biosphere Reserve. Sweden Ecol Soc 12:28. https://doi.org/10.5751/ES-01976-120128

O'Neill SJ, Handmer J (2012) Responding to bushfire risk: the need for transformative adaptation. Environ Res Lett 7:014018. https://doi.org/10.1088/1748-9326/7/1/014018

Otero I (2011a) Matadepera, Barcelona Metropolitan Region, Spain: Peasant memory for wildfire prevention Recovery of social-ecological memory for building resilience to wildfires (Policy Report). United Cities and Local Governments Committee on Social Inclusion and Participative Democracy. Available at https://serval.unil.ch/fr/notice/serval:BIB_B750744BDD86

Otero I (2011b) Set anys de recerca acció a Matadepera i al Parc Natural de Sant Llorenç del Munt i l'Obac. In: VII Monografies de Sant Llorenç del Munt i l'Obac. Diputació de Barcelona, Barcelona, pp 22–27

Otero I (2010) The rural-urban socioecological transformation of Mediterranean mountain areas under global change. Local studies in Olzinelles and Matadepera (Barcelona Metropolitan Region) (PhD Thesis). Autonomous University of Barcelona, Spain

Otero I, Boada M (2007) Perspectiva socioecológica en el análisis de áreas naturales protegidas en la Región Metropolitana de Barcelona (España). Gest Ambiente 10:125–130

Otero I, Boada M, Badia A et al (2008) El canvi global a la conca de la Tordera. In: Els sistemes socioecològics de la conca de la Tordera. Institució Catalana d'Història Natural, pp 95–129

Otero I, Boada M, Tàbara JD (2013) Social–ecological heritage and the conservation of Mediterranean landscapes under global change. A case study in Olzinelles (Catalonia). Land Use Policy 30:25–37. https://doi.org/10.1016/j.landusepol.2012.02.005

Otero I, Castellnou M, González I et al (2018) Democratizing wildfire strategies. Do you realize what it means? Insights from a participatory process in the Montseny region (Catalonia, Spain). PLoS One 13:e0204806. https://doi.org/10.1371/journal.pone.0204806

Otero I, Darbellay F, Reynard E et al (2020a) Designing inter- and transdisciplinary research on mountains: what place for the unexpected? Mt Res Dev 40:D10–D20. https://doi.org/10.1659/MRD-JOURNAL-D-20-00036.1
Otero I, Gamboa G, Bueno C et al (2020b) Democratitzar la presa de decisions sobre els incendis forestals i adaptar-se al canvi climàtic. Un mètode aplicat a la reserva de la biosfera del Montseny. Quad Agrar 48:7–42. https://doi.org/10.2436/20.1503.01.109
Otero I, Kallis G, Aguilar R et al (2011) Water scarcity, social power and the production of an elite suburb: the political ecology of water in Matadepera, Catalonia. Ecol Econ 70:1297–1308. https://doi.org/10.1016/j.ecolecon.2009.09.011
Otero I, Nielsen JØ (2017) Coexisting with wildfire? Achievements and challenges for a radical social-ecological transformation in Catalonia (Spain). Geoforum 85:234–246. https://doi.org/10.1016/j.geoforum.2017.07.020
Pahisa F (2021) Collserola podria cremar en sis hores, segons el cap del GRAF. Betevé, 15 June. Available at https://beteve.cat/medi-ambient/collserola-podria-cremar-sis-hores-segons-cap-forestals-graf/. Accessed 13 Aug 2021
Pais S, Aquilué N, Campos J et al (2020) Mountain farmland protection and fire-smart management jointly reduce fire hazard and enhance biodiversity and carbon sequestration. Ecosyst Serv 44:101143. https://doi.org/10.1016/j.ecoser.2020.101143
Piñol J, Beven K, Viegas DX (2005) Modelling the effect of fire-exclusion and prescribed fire on wildfire size in Mediterranean ecosystems. Ecol Model 183:397–409. https://doi.org/10.1016/j.ecolmodel.2004.09.001
Piñol J, Castellnou M, Beven KJ (2007) Conditioning uncertainty in ecological models: assessing the impact of fire management strategies. Ecol Model 207:34–44. https://doi.org/10.1016/j.ecolmodel.2007.03.020
Piqué M, Castellnou M, Valor T et al (2011) Integració del risc de grans incendis forestals (GIF) en la gestió forestal: Incendis tipus i vulnerabilitat de les estructures forestals al foc de capçades, Orientacions de gestió forestal sostenible per a Catalunya (ORGEST). Centre de la Propietat Forestal. Departament d'Agricultura, Ramaderia, Pesca, Alimentació i Medi Natural. Generalitat de Catalunya
Piqué M, Domènech R (2018) Effectiveness of mechanical thinning and prescribed burning on fire behavior in *Pinus nigra* forests in NE Spain. Sci Total Environ 618:1539–1546. https://doi.org/10.1016/j.scitotenv.2017.09.316
Plana E (2011) Integració del risc d'incendis en la planificació forestal estratègica i l'ordenació del territori. Treb Soc Catalana Geogr 71–72:69–91
Plana E, Martín D, Font M et al (2015) Social factor and territorial dimension of wildfire risk management: managing societal involvement and cross-sectoral planning. Project report FIREfficient. Available at http://firefficient.ctfc.cat/wp-content/uploads/2014/03/deliverable15_Wildfire-social-assesment-and-governance.pdf
Popoveniuc B (2014) Self reflexivity. The ultimate end of knowledge. Procedia – Soc Behav Sci 163:204–213. https://doi.org/10.1016/j.sbspro.2014.12.308
Prior T, Eriksen C (2013) Wildfire preparedness, community cohesion and social–ecological systems. Glob Environ Change 23:1575–1586. https://doi.org/10.1016/j.gloenvcha.2013.09.016
Reason P, Bradbury H (2008) The SAGE handbook of action research. Participative inquiry and practice. SAGE Publications
Regos A, Aquilué N, López I et al (2016) Synergies between forest biomass extraction for bioenergy and fire suppression in Mediterranean ecosystems: insights from a storyline-and-simulation approach. Ecosystems 19:786–802. https://doi.org/10.1007/s10021-016-9968-z
Regos A, Aquilué N, Retana J et al (2014) Using unplanned fires to help suppressing future large fires in Mediterranean forests. PLoS One 9:e94906. https://doi.org/10.1371/journal.pone.0094906
Rojas E (1999) El bosc mediterrani en el segle XXI. In: Medi Ambient, Tecnologia i Cultura, vol 23. Departament de Medi Ambient de la Generalitat de Catalunya, Barcelona

Ruiz V, Otero I, Aguilar R (2008) El consum d'aigua i l'especulació urbanística a Matadepera (1931–1983). Primers resultats del projecte "Memòries d'una feixa". Terme 23:203–238
San-Miguel-Ayanz J, Durrant T, Boca R et al (2018) Forest fires in Europe, Middle East and North Africa 2017. (EUR 29318 EN). Publications Office of the European Union, Luxembourg https://doi.org/10.2760/663443
San-Miguel-Ayanz J, Moreno JM, Camia A (2013) Analysis of large fires in European Mediterranean landscapes: lessons learned and perspectives. For Ecol Manag 294:11–22. https://doi.org/10.1016/j.foreco.2012.10.050
San-Miguel-Ayanz J, Oom D, Artès T et al (2020) Forest fires in Portugal in 2017. In: Casajus Valles A, Marin Ferrer M, Poljanšek K et al (eds) Science for disaster risk management 2020: acting today, protecting tomorrow (EUR 30183 EN). Publications Office of the European Union, Luxembourg, pp 413–430. https://doi.org/10.2760/571085
Schäfer M, Lux A, Bergmann M (2020) Editorial to the special issue "Transdisciplinary sustainability research. Linking research processes and outputs to societal effects". Environ Sci Pol 107:206–210. https://doi.org/10.1016/j.envsci.2020.02.018
Schneider F, Giger M, Harari N et al (2019) Transdisciplinary co-production of knowledge and sustainability transformations: three generic mechanisms of impact generation. Environ Sci Pol 102:26–35. https://doi.org/10.1016/j.envsci.2019.08.017
Spies TA, White EM, Kline JD et al (2014) Examining fire-prone forest landscapes as coupled human and natural systems. Ecol Soc 19:9. https://doi.org/10.5751/ES-06584-190309
Tàbara D, Saurí D, Cerdan R (2003) Forest fire risk management and public participation in changing socioenvironmental conditions: a case study in a Mediterranean region. Risk Anal 23:249–260. https://doi.org/10.1111/1539-6924.00305
Thompson M, Dunn C, Calkin D (2015) Wildfires: systemic changes required. Science 350:920–920. https://doi.org/10.1126/science.350.6263.920-b
Thompson MP, MacGregor DG, Dunn CJ et al (2018) Rethinking the wildland fire management system. J For 116:382–390. https://doi.org/10.1093/jofore/fvy020
Turco M, Rosa-Cánovas JJ, Bedia J et al (2018) Exacerbated fires in Mediterranean Europe due to anthropogenic warming projected with non-stationary climate-fire models. Nat Commun 9:3821. https://doi.org/10.1038/s41467-018-06358-z
Varela E, Górriz-Mifsud E, Ruiz-Mirazo J et al (2018) Payment for targeted grazing: integrating local shepherds into wildfire prevention. Forests 9:464. https://doi.org/10.3390/f9080464
Westerling AL, Hidalgo HG, Cayan DR et al (2006) Warming and earlier spring increase Western U.S. forest wildfire activity. Science 313:940–943. https://doi.org/10.1126/science.1128834
Williams AP, Abatzoglou JT, Gershunov A et al (2019) Observed impacts of anthropogenic climate change on wildfire in California. Earths Future 7:892–910. https://doi.org/10.1029/2019EF001210

Chapter 8
Formal and Disruptive Co-production of the Climate Emergency Response: The Case of Barcelona

Mar Satorras

Abstract The climate emergency momentum has transformed urban climate politics and activism. Among the demands reinvigorated by climate movements, the need for more democratic climate governance has introduced new ways of reengaging citizens in the design and implementation of urban climate action. This chapter focuses on the co-production processes that emerged around climate emergency declarations in cities by analysing the case study of Barcelona (Spain). Using previous literature emphasising the need to politicise new forms of engagement, I identify two distinctive forms of co-producing the climate emergency response. On the one hand, the institutional co-production of the Climate Emergency Declaration by which the City Council created a board and invited stakeholders to participate in the definition. and prioritisation of actions. On the other hand, the struggle of new climate movements to speed up and contest the contents and procedures defining both the Climate Emergency Board and the resultant declaration configured a form of *co-production as resistance*. The chapter ends up by discussing the transformative potential and the implications for urban resilience derived from the coexistence of formal and disruptive forms of co-producing the responses to the climate emergency.

Keywords Barcelona · Climate activism · Climate emergency declarations · Co-production · Urban climate governance

---

M. Satorras (✉)
Internet Interdisciplinary Institute (IN3), Universitat Oberta de Catalunya (UOC), Barcelona, Spain

Barcelona Institute of Regional and Metropolitan Studies (IERMB), Universitat Autònoma de Barcelona (UAB), Bellaterra, Spain
e-mail: mgrausat@uoc.edu

I. Ruiz-Mallén et al. (eds.), *Urban Resilience to the Climate Emergency*, The Urban Book Series, https://doi.org/10.1007/978-3-031-07301-4_8

8.1 Introduction

By the end of 2021, almost 2000 local governments worldwide had declared a climate emergency by passing binding motions (CEDAMIA 2021; ICEF 2021). While many resolutions are instigated by grassroots movements and include setting up a process to develop local climate planning, there is no single path to create and operationalise these declarations (Howarth et al. 2021; Ruiz-Campillo et al. 2021). Indeed, scholars are still discussing to what extent local climate emergency declarations are incorporating urban resilience and adaptation concerns or they are essentially focused on mitigation (see initial evidence from Australia, New Zealand and the UK in Davidson et al. 2020; Howarth et al. 2021; Nissen and Cretney 2021). Yet, some common defining traits include a renewed attention by local governments on advocacy and collaboration with state and international levels of climate governance, more ambitious targets and commitments (e.g. accelerate the timing to achieve net zero carbon emissions sooner), increased transparency through regular reporting and, importantly, further involvement of citizens in the co-production and co-implementation of the climate emergency response (Gills and Morgan 2020; UNFCC 2020). Regarding the latter, the first local government in declaring a climate emergency in 2016 (the city of Darebin, in Australia) argued that a greater emphasis on community mobilisation and engagement is one of the key differences between climate emergency strategies and traditional local climate planning (Darebin City Council 2019).

Democratisation of climate action through more participatory and deliberative tools has also been a pivotal demand by new climate movements (de Moor et al. 2021). In particular, Extinction Rebellion (XR), the non-violent civil disobedience movement calling for government action to prevent climate and environmental crises, has emphasised the need to *go beyond politics* by creating citizens' assemblies to inform climate emergency policies and by generally propelling what they label as the *deliberative rebellion* (XR 2019; see also Berglund and Schmidt 2020). As they explain throughout their principles, *to develop our resilience as multiple collapses take place*, climate movements should focus on transformative changes towards, among other horizons, *a functioning democracy, where people have real agency in decision-making* (XR 2021).

Thus, the local responses to the climate emergency, promoted by both urban governments and climate movements, have become a fertile arena to experiment with new forms of engagement, contestation and co-production, potentially contributing to transformation (Kythreotis et al. 2021; Patterson et al. 2021). However, it is still unclear how and to what extent such reengagements with urban climate action mobilise action conducive to transformative urban resilience. Urban transformation broadly refers to deep and radical changes disrupting current pathways by altering existing urban structures, cultures and practices (Wolfram 2016; see more on the conceptual approaches to transformation used in the sustainability literature in Patterson et al. 2017 and Chap. 2). However, in the context of urban resilience, the conceptual differences between adaptive and transformative resilience remain undertheorised, ambiguous and even contested (Elmqvist et al. 2019; Leixnering and Höllerer 2021). For instance, while urban ecologists argue that transformation approaches aim to reduce unwanted resilience and enhance desired resilience

(Elmqvist et al. 2019), it remains unclear by whom should be (or not) desirable such resilience. Similarly, Moloney and Doyon (2021) consider transformative urban resilience as the capacity to influence, disrupt and change mainstream urban policy and planning frameworks, but fail to clearly answer what is the status quo and what means to challenge it. Such questions appear less abstract when developed by scholarship from climate/environmental justice and critical urban studies. The literature on urban climate justice has argued that to foster transformative change through planning projects that purport to be resilient questions of social justice and equity should be central and not *fall into the same planning 'traps' of privileging or protecting elite groups at the expense of disadvantaged groups* (Anguelovski et al. 2016, p.344). In this regard, scholars advocate that employing social justice as an orientating principle is vital for progressing towards climate transformations that are democratically and ethically defensible (Patterson et al. 2018), which can ultimately contribute to the rights of urban vulnerable groups (Ziervogel et al. 2017). This view is related to critiques to incremental climate adaptation approaches claiming that transformation requires addressing the root causes of climate vulnerability, many of which connect directly to underlying socioeconomic development pathways and power relations (Pelling 2011). Finally, critical urban studies frame transformation as opposed to the homogeneous neoliberal trend of urban developments to respond to climate change (Long and Rice 2019). However, postcolonial perspectives broaden this view by showing how neoliberal climate pathways likely coexist with alternative climate practices and articulate a proposal to investigate these diverse forms of urban climate action so as to explore their transformative potential while addressing their *capacity to improve living conditions for the urban majority under climate change* (Robin and Castán-Broto 2021, p.5). Thus, transformation under these approaches is entangled with social justice, but also with emancipatory agendas.

Previous research has emphasised how collaborations between municipalities and citizens around urban climate adaptation can foster equity and climate justice (Brink and Wamsler 2018) and therefore support urban changes with transformative potential. Similarly, informal and community-based climate resilience action propelled by social movements, grassroots and other bottom-up initiatives seem to be crucial to dispute the status quo and move towards transformative resilience (e.g. Cloutier et al. 2018; Seebauer et al. 2019). As recently argued by Castán-Broto and Robin (2021, p.3), *everyday, Do-It-Yourself (DIY) or informal urbanism may help to reframe climate urbanism away from the capitalist appropriation of good intentions that pervade climate change politics.* However, the literature has often approached separately these potential enablers of transformation by focusing on either government-led climate planning (e.g. Satorras et al. 2020; Ruiz-Campillo et al. 2021) or social movements and grassroots actions (e.g. MacGregor 2021; Pickerill 2021). Moreover, the incipient body of knowledge willing to understand recent local climate emergency responses (e.g. Harvey-Scholes 2019; Chou 2020; Davidson et al. 2020; Howarth et al. 2021; Kythreotis et al. 2021; Miller 2021; Nissen and Cretney 2021; Ruiz-Campillo et al. 2021) has still not critically examined their transformative potential with a focus on urban resilience.

To fill these gaps, this chapter aims to contribute to discussions about new forms of engagement and contestation appeared when addressing the climate emergency in the urban context and decipher both their implications in terms of urban resilience and their transformative potential. By approaching transformative resilience through the lenses of critical urban studies and environmental justice, I develop a case study based on the politicisation of the co-production of the Climate Emergency Declaration in the city of Barcelona. Politicisation implies a return of the political which remains crucial amidst contemporary depoliticisation processes, characterised by the removal from the public terrain of the political dimension, *understood as a space of contestation and agonistic engagement*, and its replacement by politics, *understood as technocratic mechanisms and consensual procedures that operate within an unquestioned framework of representative democracy, free market economics and cosmopolitan liberalism* (Wilson and Swyngedouw 2014, p.6). By discussing how and to what extent such co-production processes may (re)introduce *the political* and therefore question and dispute current urban political and economic structures and cultures, this chapter pretends to inform the debates about the transformative potential of these new climate activism tactics and forms to govern with and for the people that have emerged in cities with the aim to address the climate emergency and improve urban resilience.

Co-production entails the involvement of actors – beyond governments and experts – in public services delivery, knowledge production or policymaking (Voorberg et al. 2015). In the urban resilience literature, knowledge co-production with multiple actors has been considered as *a process to invite, facilitate and enable locally informed and globally related meanings of urban resilience and sustainability* (Elmqvist et al. 2019, p.271). However, critical analyses have identified at least seven potential pitfalls that may emerge when co-production is implemented in practice: the deliberate rejection of responsibility, failing accountability, rising transaction costs, loss of democracy, reinforced inequalities, implicit demands and co-destruction (Steen et al. 2018). To further question and better acknowledge the limitations of institutional co-productive initiatives, some scholars argue that the political should be reintroduced so as to identify different and potentially competing forms of co-production. In this regard, Flinders and Wood (2019) develop a new taxonomy which distinguishes co-production initiatives set up by the state (i.e. *institutional co-production*), from co-production practices empowering citizens to resist top-down and often market-driven policies and to demand a stronger role in the codesign of public policy (i.e. *co-production as resistance*). According to them, institutional co-production occurs in *invited spaces*, i.e. when the state invites citizens to co-produce in a range of forums, while co-production as resistance or disruptive co-production materialises in *insisted spaces*, i.e. when citizens have to insist they want to be heard in arenas for public debate (Carson 2008; Flinders and Wood 2019). Inspired by this distinction, I develop a critical view on the co-production processes that emerged around the Climate Emergency Declaration in Barcelona (2020), which supports my discussion about their capacity to mobilise transformative action with a focus on urban climate resilience.

After this introduction, I lay out the methodology used to develop this research. The results are narrated in two sections: the first addressing the formal co-production process of the Climate Emergency Declaration led by the city government of Barcelona and the second delving into the confrontational co-production practices used by local climate activism to speed up and contest the contents and procedures defining the Climate Emergency Board and the resultant declaration. Then, I discuss and compare the outputs of both processes by looking at their transformative potential and the implications for urban climate resilience. Finally, I conclude the chapter by arguing that the coexistence of formal and disruptive forms of co-producing the responses to the climate emergency can enable more transformative urban resilience policies and, in overall, instigate climate action that spur change.

8.2 Methodology

This chapter is mainly based on 12 semi-structured interviews conducted to 4 public agencies and 8 community organisations, to understand their perceptions and experiences about the co-production of the Barcelona Climate Emergency Declaration (see Table 8.1). The interview covered several topics regarding climate and urban resilience governance. In this chapter, I focus on four specific questions addressing (1) the novelties brought by the climate emergency (as compared with previous efforts and strategies against climate change in the city), (2) the role their organisation (did or did not) played in the Barcelona Climate Emergency Declaration, (3) how they experienced and valued the formal engagement process to co-produce it (i.e. the Climate Emergency Board) and (4) what was the added value of the engagement process and what were the challenges faced. In the case of social movements, questions about their actions, tactics and mobilisations were also included. Interviews were conducted from June to November 2020 and took place online due to COVID-19 restrictions (in either Catalan or Spanish). Audio records were transcribed and qualitatively analysed by codifying them through Atlas.ti. Informed consent was obtained from all interviewees.

Data from interviews was integrated with observation notes taken by the author during institutional and activist events, the information provided by the digital participation platform (Decidim) supporting the Climate Emergency co-production process (see next section) and a review of grey literature on the development of public and communitarian climate action in the city including planning documents, press releases, newspapers, and blog posts.

Finally, as part of the evaluation of the co-production process, a survey was conducted among the participants of the Climate Emergency Board from December 2019 to January 2020. Online questionnaires were sent by email to all organisations who were known to have participated in one or more workshops of the process ($n = 198$), except those from the public administration ($n = 27$). Surveys were only filled by 29 organisations (17%), and therefore the response rate was too low to provide representative results. Thus, to secure valid and reliable results in this research, I use the data from this survey to complement the interviews, the observations and the review of grey literature and not as the main source of data. The survey contained a range of questions about respondent's perception of the co-production process, including their motives for participation in the Climate Emergency Board,

Table 8.1 Stakeholders interviewed from the Barcelona case study

Stakeholders interviewed	Description	Interview code
Barcelona Climate Office	*Office responsible to elaborate, support the implementation and follow up climate planning in the city of Barcelona*	Interview #1
Barcelona Urban Resilience Department	*Department coordinating urban resilience actions across the Barcelona City Council. It also received funding and support from the 100 Resilient Cities programme (funded until 2020 by the Rockefeller Foundation)*	Interview #2
Metropolitan Climate Emergency and Environmental Education Service	*Unit responsible for climate change, energy and environmental education in the Metropolitan Area of Barcelona*	Interview #3
Catalan Climate Change Office (OCCC)	*Office responsible to promote climate change strategies, planning and projects in Catalonia*	Interview #4
Federation of Barcelona Neighbourhood Associations (FAVB)	*The federation brings together more than a hundred associations of residents of the neighbourhoods of Barcelona. It aims to improve the quality of life of all citizens so as to support equality, solidarity and coexistence*	Interview #5
Fridays for Future (FFF) Barcelona	*Local node of climate mobilisations such as #Friday4Future and #Youth4Climate led by students*	Interview #6
Rebel·lió o Extinció (XR) Barcelona	*Local node of the international movement Extinction Rebellion which uses non-violent disobedience to struggle against massive ecological extinction and collapse*	Interview #7
ClimAcció	*Local collective that works with artivism and non-violent struggle against the climate crisis*	Interview #8
Enginyeria Sense Fronteres (ESF)	*NGO for development cooperation constituted by engineers, teachers and students, which among other goals aims to guarantee universal access to basic services and promote technology oriented towards human development*	Interview #9
Sustainability Secretariat – CCOO	*Secretariat of sustainability policies from the largest Spanish trade union confederation which includes different territorial and sectorial organisations*	Interview #10
Arquitectura i Sostenibilitat – COAC	*Group from the Architects' Professional Association in Catalonia (CoAC) which aims to mainstream and generate debate around sustainability (particularly within architects) and to develop an environmentally friendly architecture*	Interview #11
Can Masdeu	*Squatted farm in the hills of Barcelona which among other actions offer their space to facilitate encounters among climate activists and organise climate-related activities and debates. Members are also experimenting with climate adaptation by recovering and improving the farm water infrastructure*	Interview #12

their evaluation of the outcomes of the co-production process as well as their role during the process.

8.3 The Formal Co-production of the Barcelona Climate Emergency Declaration

In May 2019, Spanish local elections left a disputed result to govern the city of Barcelona. After a few weeks of negotiations, the anti-austerity party *Barcelona en Comú* led by Ada Colau kept the mayor's office through a coalition with the socialist party, PSC (ElPais 15/6/ 2019). Although the electoral programme from both political parties proposed to implement the Climate Plan approved the previous year and did not even mention the notion of the *climate emergency* (see Barcelona en Comú 2019, p.86, and PSC 2019, p.48), the local government established that declaring a climate emergency was one of the key priorities for the new 4-year term (Barcelona City Council 2019a; see more on the city environmental governance in Chap. 4).

Importantly, the content of the Barcelona Climate Emergency Declaration was to be co-produced through a board of organisations considered as a *steering group* that *will set out the ambition, scope and concrete measures to deal effectively with the climate emergency and will monitor and evaluate them. It will also be the space where the City Council will be accountable for the progress and fulfilment of the commitments made* (own translation, Barcelona City Council 2019b, p.2). The co-production tools, similar to the latest Climate Plan co-production process (see Satorras et al. 2020), consisted of four face-to-face thematic workshops (see Table 8.2) as well as a digital platform where proposals could be uploaded, commented and valued.[1] While the workshops were designed to involve organisations, the digital platform was also opened to lay citizens (Interview #1). Workshops were organised by pre-defined topics which were modified over the process from the initial proposal (Barcelona City Council 2019b) by including new topics (i.e. water and coastal management, climate justice and culture, health) and issues traditionally not integrated in previous climate plans in principle due to the limited powers of local governments (i.e. Port and Airport, intercity mobility and climate taxation). In Sect. 8.5, I will discuss the possible reasons and consequences derived from this ambitious and broadened perspective.

In total, 198 different organisations were involved in the co-production process. Previous co-production processes to elaborate climate plans and strategies had engaged from 70 to 140 organisations (Satorras 2021), and therefore the momentum raised by the climate emergency in 2019 could probably explain this larger engagement. In fact, the organisers recognised that thanks to the climate emergency, *people who had never come to previous (climate-related) participation processes* was also involved (Interview #1). Most participants in the Climate Emergency Board were NGOs or foundations (31%) and private sector entities (25%), which were followed by public sector entities (14%), professional and business associations (11%),

[1] See the process of the Barcelona Climate Emergency Board in the digital platform *Decidim*: https://www.decidim.barcelona/assemblies/emergencia-climatica.

Table 8.2 Description of the workshops and participants at the Barcelona Climate Emergency Board co-production process

Workshop	Date	Topics discussed	Number of participant organisations	Number of people participating
1st	3 October 2019	1. Water and coastal management 2. Urban greening 3. Renewable energy 4. Food sovereignty	133	174
2nd	17 October 2019	1. Vulnerable people to climate change 2. Urban mobility 3. Waste management 4. Climate justice and culture	92	107
3rd	7 November 2019	1. Emissions and health 2. Sustainable urbanism 3. Building rehabilitation 4. Circular economy and responsible consumption	77	104
4th	28 November 2019	1. Port and Airport 2. Intercity mobility 3. Climate taxation	59	81

Source: own elaboration with data from the *Decidim* platform

universities and research centres (7%), primary and secondary schools (7%) as well as some major trade unions (2%), political parties (2%) and press (2%) (see Fig. 8.1).

Three quarters of the involved organisations only attended one or two of the four workshops (54% and 21%, respectively). While most of these sporadic participants attended the first workshop, some organisations preferred to engage in specific workshops targeting their interests. For instance, the platform ZerØport (a local degrowth-inspired movement targeting the Barcelona Port and Airport) only attended the last workshop which precisely discussed about these infrastructures, while the RACC Foundation (the largest club of car drivers in Spain) only participated in the second workshop which tackled urban mobility issues (see the topics in Table 8.2). The remaining organisations engaged in three or all the four workshops (15% and 10%, respectively). It is important to notice that a third of the organisations able to follow the four workshops and therefore the whole process were public institutions such as municipal agencies, natural park managers and metropolitan authorities (see orange bars in Fig. 8.1). These different capacities to participate in the process were noticed by some interviewees from voluntary associations or NGOs who contended that *the people who participate is mostly people from institutions, people who have a minute to be in these meetings, and we are the only volunteer participants* (Interview #5).

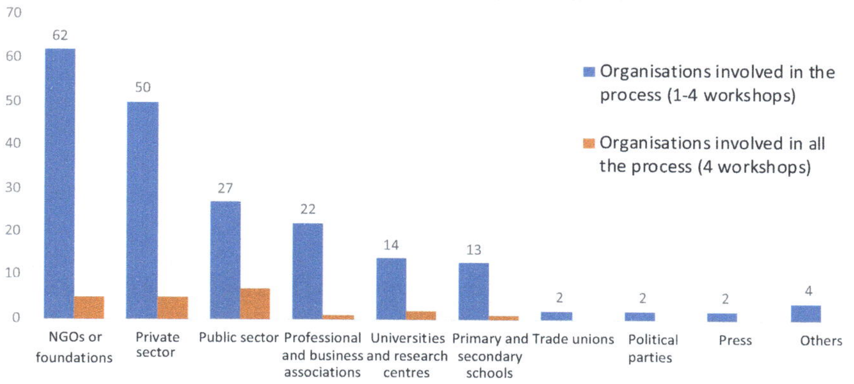

Fig. 8.1 Organisations involved in the Barcelona Climate Emergency Board, by typology and degree of participation. (Source: own elaboration with data from *Decidim* platform)

The informants answering the survey were predominantly from these highly involved organisations.[2] These informants valued positively the face-to-face sessions of the Climate Emergency Board (3.8/5) and the supporting digital platform (3.5/5). They particularly appreciated the facilitation of the workshops, the human and economic resources invested by the City Council in this process and the degree of participation of the organisations (all valued at 3.8/5), while the items less valued were the depth of the debates (2.8/5) and the degree of lay citizens' participation (2.9/5). Several participants interviewed also pointed at this limited time to explore in depth each issue (Interviews #5, #9, #10), while civil servants expressed the challenges they perceive – such as the large resources that should be allocated – to organise massive climate-related participatory processes actively involving lay citizens (Interviews #1, #3). In fact, both challenges – low participation from lay citizens and depth of the debates – were concerns already raised by participants from the last climate policymaking co-production process in the city (Satorras et al. 2020).

In total, 62 proposals with potential actions to be included in the Climate Emergency Declaration were collected through the workshops and the platform. The final policy document organised almost a hundred climate actions under seven structural changes (i.e. across urbanism, mobility and infrastructure, energy, economy, consumption and waste, food and culture and education) and envisioned two lines for building urban climate resilience focusing on health, well-being and environmental quality and water (Barcelona City Council 2020a). The resulting Climate Emergency Declaration was presented on 15 January 2020 by the Mayor as not

[2] The average participation in the Board was 2.9 workshops for survey respondents ($n = 29$), while the overall involvement was 1.8 workshops ($n = 198$). In other words, the sample is biased towards motivated participants: two thirds of the organisations surveyed had participated to three or four workshops (37% and 30%, respectively), while only one third of the respondents had participated to one or two workshops (11% and 22%, respectively).

merely *a rhetoric decree, signature* or *empty declaration*, but *a real process with concrete measures and commitment to mark a turning point in the struggle against the climate emergency* (Colau 2020). In her speech, the Deputy Mayor for Environment also highlighted the collaborative dimension of the process by noting that it was *not an ordinary emergency declaration, but the result of the work that we have all done during these last 4 months* (Sanz 2020a). The surveyed participants valued positively the overall process, taking into account both the process and the outputs (3.4/5); and, in particular, they appreciated the social sensitivity of the resulting actions (3.3/5) and their utility to address the climate emergency in the city (3.2/5). However, the item regarding the outputs of the process least valued was the political commitment to implement and therefore put into practice co-produced actions (2.7/5), a scepticism that I will further examine in the next section.

8.4 The Disruptive Co-production of the City Climate Emergency Response

The disruptive co-production of the climate emergency response was developed in parallel to the formal engagement process and therefore reproduced similar chronologies between the summer of 2019 and early 2020. Returning to July 2019, the same day the Mayor announced the constitution of the Board to define the contents of the Climate Emergency Declaration, climate movements integrated by XR Barcelona, Fridays for Future Barcelona, Families for Future Barcelona, ClimAcció and the Movement for Climate Justice sent a letter to the Mayor. In this open letter, they noted that the new term was *critical to tackle the climate and environmental emergency* and proposed seven key policies to address *seriously* climate change (XR Barcelona 2019a, Interviews #6, 7, 8). While these urgent policies mainly aimed to reduce greenhouse emissions (e.g. a moratorium to the expansion of fossil fuel infrastructures such as the Airport or the Port, an urban road toll to regulate traffic entering the city or the increase of renewable energy generation), they also included two strategies contributing to urban climate resilience (i.e. the promotion of food sovereignty and the implementation of an effective climate communication and awareness plan) (XR Barcelona 2019a). They concluded the letter by proposing to discuss these demands in a meeting after next global climate mobilisations.

The mobilisations materialised in the form of the third Global Week for Future planned for September 2019, in the context of the United Nations Climate Action Summit, which took place in more than 150 countries and mobilised above six million people (de Moor et al. 2021). In the case of Barcelona, the climate movements joined the global climate strike and organised a week of climate-related events which ended with the largest climate demonstration in the city to date. In the manifesto of the convening entity, Fridays for Future, the City Council was urged again to implement the seven measures proposed (Fridays for Future Barcelona 2019). After the mass demonstration, the five organisations integrating the climate movement met with the Mayor and two councillors. During the meeting they insisted in the urgency to redress local climate policies which they deem to be insufficient (Interviews #6, 7, 8). All in all, they perceived *readiness, but not enough concretion* from the side of the city government (XR Barcelona 2019b).

Importantly, the organisations integrating the climate movement also raised their concerns regarding the Climate Emergency Board and formal co-production process that was due to start. In particular, they considered that the declaration of the state of climate emergency required *priority and determination about the specific government policies that cannot be delegated to a co-production space with diverse actors in four sessions, following the same sterile participative outline of other boards that tend to hide conflicts under false consensus* (XR Barcelona 2019b). In this line, they insisted that the Board could be useful to guarantee transparency and accountability of governmental action, but *not as a substitute for it* (Idem). In other words, they seem to disapprove the so-called shift from *government* to *governance* by pointing at potential democratic shortfalls such as false consensus also noted by some scholarship (Swyngedouw 2005; Allmendinger and Haughton 2012), but particularly by criticising the invitation of powerful lobbies. As argued by one of the organisations from the climate movements, *we did not like the fact that the City Council, the social groups and the private companies were on the same board. We preferred first to speak with the City Council to really put the climate emergency at the centre and speak freely. And later, with this information, the City Council can talk and negotiate with the companies. But, in our opinion, to put ourselves in front of the companies is unfair* (Interview #8). Thus, their proposal to rethink this participative space revolved around the participants included and also the Climate Emergency Board purposes. On the one hand, they argued that *business lobbies* should be excluded from the Board since they were representing private interests and the fossil fuel economy and were therefore *responsible of the problems of Barcelona* rather than potential solvers (Interview #6). Instead, climate activists proposed to target participants representing the general interest such as *social agents and independent scientific experts* (XR Barcelona 2019b). On the other hand, they considered that the ambition of the Board should lie on the monitoring and improvement of the implementation of the actions and the outputs of the Climate Emergency Declaration rather than focusing on the co-creation of the declaration.

Local climate movement's demands were not formally incorporated in the design of the Board goals and process, but, still, the climate movement and the network of organisations closely collaborating with them participated in the workshops to define the Climate Emergency Declaration, although they were critically involved. For instance, in the first workshop, they asked the local government to avoid long time horizons and instead present the main measures planned to address the climate emergency during the 2019–2023 term in detail (e.g. scale, calendar and quantitative assessment of the contribution of each action to mitigate or adapt climate change) (XR Barcelona 2019b). They also tried to struggle against proposals from the above-mentioned lobbies. For instance, during the debate on water, the activists for public water – well aligned with the local climate movement position – questioned proposals from the private water operator. These activists defended that the current private-oriented water governance model should be replaced for public government supported by collaborative approaches to face climate change (Interview #9). Some climate activists were precisely involved to counterbalance the possible influence from big private companies: *when we saw that Airbnb and other multinationals were part of the Board, we registered to serve as a citizen counter-power and to push for measures other than greenwashing* (Interview #8). Using more confrontative tactics, during the last workshop, the local climate movements made their demands regarding the Port and the Airport visible through a performance (see Fig. 8.2; XR Barcelona 2019c). While they acknowledged the limited competences of the City Council, local activists assured that this action aimed to put attention to the Port and the Airport and *avoid only talking and not acting* (LaVanguardia 2019).[3]

Finally, in January 2020, the organisations integrating the local climate movement attended the official presentation of the Barcelona Climate Emergency Declaration, although they warned that they *will analyse in detail all the measures of the plan* and insisted on the need *to create an independent committee to monitor them* with citizens and scientists (XR Barcelona 2020a). In the public event, a well-known invited activist from Fridays for Future reminded that *social and environmental justice movements will be watching (…) and won't allow any delaying tactics occurring for either economic or electoral interests* (Barricarte 2020).

Fig. 8.2 Performance from the local climate movement during the last workshop of the Climate Emergency Board. (Source: Mar Satorras (28 November 2019))

[3] This critique to empty promises and passive politicians for their inaction over climate change has long been present in climate activism discourses specially since the failed 2009 Copenhagen Climate Summit (Chatterton et al. 2013; Bäckstrand and Lövbrand 2019) and has been recently popularized through the dismissal of new climate and green recovery pledges as *blah, blah, blah* after the viral speech from Greta Thunberg in the 2021 Youth4Climate Summit.

Accompanying her warning, at the front row, another activist exhibited a banner with a drawn big eye symbolising their willingness to continue to critically scrutinise local policymakers' climate action.

In the next section, I will briefly present the outcomes of the disruptive and formal co-production processes underlying the development of the Barcelona Climate Emergency Declaration so as to discuss their implications in terms of urban climate resilience and potential for enabling transformation in resilience and climate policy and practice.

8.5 The Results of Co-production(s) and the Implications for Urban Transformative Resilience

To better understand the potential of climate emergency declarations, initial empirical analyses among early local government adopters are trying to figure out the differences between these new emergency plans and previous local climate action and planning (see, for instance, Davidson et al. 2020; Howarth et al. 2021; Nissen and Cretney 2021; Ruiz-Campillo et al. 2021). In the case of Barcelona, the resulting Climate Emergency Declaration significantly increased the ambition from previous climate planning (i.e. the 2018 Barcelona Climate Plan) in terms of mitigation targets: from a 45% drop of GHG emissions by 2030 compared to 2005 to reducing 50% of emissions by 2030 compared to 1992 (Barcelona City Council 2020a). Nonetheless, as one of the civil servants who was highly involved in both policies recognised, *in the Declaration there are not many new or different actions from those in the Climate Plan* (Interview #2). In fact, the coordinator of both climate schemes at the City Council explained that *we have merged them* because *we have increased the ambition (as compared with the Climate Plan), but the actions are more or less the same* and *having two repetitive documents does not facilitate implementation or follow-up* (Interview #1).

Thus, according to the local government adopters' views, the main outcome of the Climate Emergency Declaration (Barcelona City Council 2020a) was an increased ambition by the City Council mostly regarding the mitigation target. However, this change regarding the proportion of emission reduction by 2030 and the baseline year to calculate it was not explicitly discussed in the formal co-production process. It is possible that such increased ambition was politically motivated for the climate movements and mobilisations. A civil servant from another public institution reflected on the role of these new climate movements by highlighting that *the Climate Emergency Declaration changed at the request of these movements, although some (city council) technicians did not agree. The Climate Plan was a very technical plan elaborated during long time and suddenly incorporating certain modifications... (was challenging). However, in the end, politics is also about that* (Interview #3). According to this interviewee, the scope of these changes was not limited to the carbon dioxide (CO_2) target, but also probably included other issues addressed in the declaration such as the Airport, the Port, air quality and the touristic sector (Interview #3).

Therefore, what I refer here as the disruptive co-production process seems to be key to incorporate radical demands from local organisations integrating the climate movement which potentially challenged powerful interests and long-established lobbies in Barcelona. In fact, when the climate movement assessed the final Climate Emergency Declaration presented by the City Council, they argued that their major achievement was the consideration of the Barcelona Port and Airport within the scope of the declaration since they were *highly emitting sectors that were not included in the previous Climate Plan* (XR Barcelona 2020b, p.1).

Despite the possible reluctances by some civil servants, at the political level, local governors seemed aligned and comfortable with these demands surpassing traditional climate governance in the city and, in particular, the previous Climate Plan. This change in the political agenda was illustrated the day before presenting the Climate Emergency Declaration, when the Barcelona Deputy Mayor for Environment and the Climate Emergency Councillor gave a press conference to anticipate the climate actions regarding large-scale infrastructures (Barcelona City Council 2020b). The conference aimed to present a technical report assessing the Port and Airport direct and indirect emissions using novel methodologies and to compare them with the overall city emissions. According to this report, both infrastructures generated four times more CO_2 than the overall city (Barcelona Regional 2020). Thus, the Deputy Mayor for Environment argued that *although we don't have power over these infrastructures located in our city (…), the Port and the Airport cannot stay at the edge of the declaration* (Sanz 2020b). During the following days, the Port, the Airport and several media harshly criticised the methodology used by the report (e.g. La Vanguardia 2020; El Economista 2020). Moreover, as the Port press release noted, they urged *a rectification by the City Council because it is questioning an activity that generates 41,000 jobs and 1.7% of Catalonia GDP* (Port de Barcelona 2020, p.2).

Although the virulent reactions from these powerful actors and the COVID-19 crisis had rapidly frozen any regulatory action on these strategic infrastructures, I consider it opened a novel debate about the scope and the distribution of responsibilities within local climate policies with potential to transform urban climate governance. In other words, the dispute (more rhetorical than having immediate tangible effects) emerged because disruptive co-production generated a new political arena to debate local climate movement's demands for radical transformation and to contest approaches based on incremental, market-driven and techno-managerial climate action (Swyngedouw 2010; Pelling 2011). Moreover, as argued by Nissen and Cretney (2021), it is possible that emergency declarations and the processes to promote, create and implement these declarations may inspire more radical changes afterwards. In the case of Barcelona, some of these debates, political contestations and alliances re-emerged and accelerated during 2021 after proposals to further expand the Barcelona Airport (ACN 2021; Meo et al. 2021).

However, disruptive co-production tactics mostly failed to instigate radical changes concerning the adaptation and resilience to local manifestations of climate change (i.e. heatwaves, floods, water scarcity and coastal degradation; see Barcelona 2018). The root causes of climate vulnerability were poorly included within local activist demands and public debates; and calls for climate justice remained more global than locally grounded. Instead, the formal co-production process seemed to be more effective at incorporating novel actions in the Climate Emergency Declaration supporting climate resilience in key sectors often overlooked by previous climate planning such as public health and food sovereignty. For instance, one of the big changes set out by the declaration was dedicated to the food model and incorporated nine actions on food sovereignty mainly shaped by the ideas and proposals from local organisations working on agroecology (Interview #1; Barcelona 2020, p.23). Thus, as an activist involved in both agroecology and community-based climate resilience reflected, *one thing is to push (the administration) for an agenda and another one is making policies with the administration. They are different things, but they can coexist and result in different outcomes* (Interview #12).

8.6 Concluding Remarks

Scholars argue that we live *in a historical moment which prioritises the urban as a site for climate action* (Castán-Broto and Robin 2021, p.2). A key debate within critical urban studies scholarship is whether, how and to what extent urban responses to climate change challenge the dominant development framework or remain another manifestation of the neoliberal urban project (Robin and Castán-Broto 2021). In this regard, recent climate emergency declarations established by multiple cities after mass mobilisations have started to be examined so as to understand their added value and differences – if any – with former climate action and planning (see Davidson et al. 2020; Howarth et al. 2021; Nissen and Cretney 2021; Ruiz-Campillo et al. 2021).

In this chapter, I have closely examined the political and activist processes behind the co-production of the Barcelona Climate Emergency Declaration, so as to discuss its potential for transformation with a particular focus on urban climate resilience. Inspired by previous literature emphasising the need to politicise new forms of engagement in general and co-production in particular (see Carson 2008; Flinders and Wood 2019), I have narrated the stories of two distinctive, but interlinked, forms of co-producing the climate emergency response. On the one hand, the formal co-production of the climate emergency action plan by which the City Council invited stakeholders to participate in a board to define and prioritise actions. On the other hand, the disruptive co-production by which the local climate movement aimed to speed up and contest the contents and procedures defining the Climate Emergency Board and the resultant Climate Emergency Declaration.

The discussion of the outcomes of both co-production processes has shed light on their respective added value and transformative potential in terms of urban climate resilience. The disruptive co-production was able to incorporate radical demands from the local climate movement into the Climate Emergency Declaration which were not traditionally considered in Barcelona climate planning. For instance, the need to focus climate action on strategic (and powerful) infrastructures and sectors (e.g. Port, Airport, tourism), even if they were beyond the powers of the City Council, was included in both the declaration and the local government's agenda. As I argue, this process enabled the emergence of a new public debate about the scope and the distribution of responsibilities within local climate policies with potential to transform urban climate governance paradigms in the city. However, my results also indicate that disruptive co-production tactics mostly failed to instigate radical changes concerning the vulnerability and resilience to local climate change effects. Instead, the formal co-production seemed to be more effective at incorporating novel actions in the Climate Emergency Declaration supporting climate resilience in key sectors often overlooked by previous climate planning such as public health and food sovereignty. This is why I conclude this chapter by pointing at the value that the coexistence of formal and disruptive forms of co-producing the responses to the climate emergency has to enable transformation of urban resilience policies and well-established urban climate governance paradigms.

Acknowledgements I would like to particularly thank all organisations, activists and public workers interviewed for this research for the time and critical insights provided. I also appreciate the support received by the Barcelona+Sostenible Secretary to design and send the online survey about the Climate Emergency Board. I am also grateful to the colleagues and co-editors of this book, I. Ruiz-Mallén and H. March, for their helpful comments provided to earlier versions of this manuscript. The views, opinions and assumptions expressed in this chapter are those of the author and do not necessarily reflect those of the city government or any agency thereof. This research has been funded by the Spanish Research Agency through a 'Juan de la Cierva – Formación' research fellowship (FJCI-2017-31723) and the Plan Nacional RESCITIES (PGC2018-100996-AI00(MCIU/AEI/FEDER, UE)).

References

ACN (2021) The complex debate over expanding Barcelona airport, 10th July 2021. https://www.catalannews.com/business/item/the-complex-debate-over-expanding-barcelona-airport

Allmendinger P, Haughton G (2012) Post-political spatial planning in England: a crisis of consensus? Trans Inst Br Geogr 37:89–103. https://doi.org/10.1111/j.1475-5661.2011.00468.x

Anguelovski I, Shi L, Chu E, Gallagher D, Goh K, Lamb Z, Reeve K, Teicher H (2016) Equity impacts of urban land use planning for climate adaptation: critical perspectives from the global North and South. J Plan Educ Res 36(3):333–348. https://doi.org/10.1177/0739456X16645166

Bäckstrand K, Lövbrand E (2019) The road to Paris: contending climate governance discourses in the post-Copenhagen era. J Environ Policy Plan 21(5):519–532. https://doi.org/10.1080/1523908X.2016.1150777

Barcelona City Council (2019a) Press release: Barcelona decretarà l'Emergència Climàtica l'1 de gener de 2020, 10th July 2019. https://ajuntament.barcelona.cat/premsa/2019/07/10/barcelona-decretara-lemergencia-climatica-l1-de-gener-de-2020/ Accessed 26 Jul 2021

Barcelona City Council (2019b) Emergència climàtica: Compromís i crida a l'acció. https://www.barcelona.cat/barcelona-pel-clima/sites/default/files/documents/emergencia_climatica._compromisos_i_crida_a_laccio.pdf Accessed 26 Jul 2021

Barcelona City Council (2020a) This is not a drill: Barcelona climate emergency declaration. https://www.barcelona.cat/emergenciaclimatica/sites/default/files/2020-07/Climate_Emergency_Declaration_en.pdf. Accessed 26 Jul 2021
Barcelona City Council (2020b) Press Conference: Sanz i Badia expliquen iniciatives relacionades amb les infraestructures i l'emergència climàtica, 14th January 2020. https://ajuntament.barcelona.cat/premsa/2020/01/13/sanz-i-badia-expliquen-iniciatives-relacionades-amb-les-infraestructures-i-lemergencia-climatica/
Barcelona en Comú (2019) Plans per una Barcelona Futura: Programa electoral 2019–2023. https://barcelonaencomu.cat/sites/default/files/document/bcomu_programa_.pdf
Barcelona Regional (2020) Estimació de les emissions dels sectors marítim i aeronàutic vinculades al Port i Aeroport de Barcelona. https://www.bcnregional.com/ca/projects/estimacio-de-les-emissions-dels-sectors-maritim-i-aeronautic-vinculades-al-port-i-aeroport-de-barcelona/
Barricarte G (2020) Intervention of Gemma Barricarte, member of Fridays for Future Barcelona and representative of the network of Barcelona climate movements. Barcelona Climate Emergency Declaration, 15th January 2020, Barcelona City Hall, Barcelona. https://youtu.be/rfW-_zIzVjA (minute 1:04:00)
Berglund O, Schmidt D (2020) Reimagining democracy. In: Extinction rebellion and climate change activism. Palgrave Macmillan, Cham. https://doi.org/10.1007/978-3-030-48359-3_5
Brink E, Wamsler C (2018) Collaborative governance for climate change adaptation: mapping citizen–municipality interactions. Environ Policy Gov 28(2):82–97. https://doi.org/10.1002/eet.1795
Carson L (2008) Creating democratic surplus through citizens' assemblies. J Publ Delib 4(1):art5. https://doi.org/10.16997/jdd.64
Castán Broto V, Robin E (2021) Climate urbanism as critical urban theory. Urban Geogr 42(6):715–720. https://doi.org/10.1080/02723638.2020.1850617
CEDAMIA (2021) Global database of Climate Emergency Declaration jurisdictions. https://www.cedamia.org/global-ced-maps/ Accessed 26 Jul 2021
Chatterton P, Featherston D, Routledge P (2013) Articulating climate justice in Copenhagen: antagonism, the commons and solidarity. Antipode 45(3):602–620. https://doi.org/10.1111/j.1467-8330.2012.01025.x
Chou M (2020) Australian local governments and climate emergency declarations: reviewing local government practice. Aust J Public Adm 80(3):613–623. https://doi.org/10.1111/1467-8500.12451
Cloutier G, Papin M, Bizier C (2018) Do-it-yourself (DIY) adaptation: civic initiatives as drivers to address climate change at the urban scale. Cities 74:284–291. https://doi.org/10.1016/j.cities.2017.12.018
Colau A (2020) Intervention of Ada Colau, Mayor of Barcelona. Barcelona Climate Emergency Declaration, 15th January 2020, Barcelona City Hall. https://youtu.be/rfW-_zIzVjA (minute 8:00)
Darebin City Council (2019) Resource for effective local government climate emergency response: Darebin Council's Climate Emergency Journey. Available at: http://www.darebin.vic.gov.au/-/media/cityofdarebin/Files/Darebin-Living/CaringfortheEnvironment/EnergyandClimate/ClimateChange/DCCClimateEmergencyCouncilsResource.ashx Accessed 15 Jul 2021
Davidson K, Briggs J, Nolan E, Bush J, Hakansson I, Moloney S (2020) The making of a climate emergency response: examining the attributes of climate emergency plans. Urban Clim 33:100666. https://doi.org/10.1016/j.uclim.2020.100666
de Moor J, De Vydt M, Uba K, Wahlström M (2021) New kids on the block: taking stock of the recent cycle of climate activism. Soc Mov Stud 20(5):619–625. https://doi.org/10.1080/14742837.2020.1836617
El Economista (2020) El Puerto y el Aeropuerto de Barcelona acusan a Colau de "manipular" sus emisiones de CO2. 15th January 2020. https://www.eleconomista.es/catalunya/noticias/10300617/01/20/El-Puerto-y-el-Aeropuerto-de-Barcelona-acusan-a-Colau-de-manipular-sus-emisiones-de-CO2.html
Elmqvist T, Andersson E, Frantzeskaki N et al (2019) Sustainability and resilience for transformation in the urban century. Nat Sustain 2(1):267–273. https://doi.org/10.1038/s41893-019-0250-1

ElPais (2019) Colau reelegida alcaldesa en una tensa investidura con los votos de PSC y Valls, 15th June 2019. https://elpais.com/ccaa/2019/06/15/catalunya/1560609352_371139.html
Flinders M, Matthew W (2019) Ethnographic insights into competing forms of co-production: a case study of the politics of street trees in a northern English city. Soc Policy Admin:1–16. https://doi.org/10.1111/spol.12484
Fridays for Future Barcelona (2019) Manifest 27S 2019. https://www.fridaysforfuturebcn.cat/manifest-27-set-2019/
Gills B, Morgan J (2020) Global Climate Emergency: after COP24, climate science, urgency, and the threat to humanity. Globalizations 17(6):885–902. https://doi.org/10.1080/14747731.2019.1669915
Harvey-Scholes C (2019) Climate emergency declarations accelerating decarbonisation? What 249 UK examples can tell us. IGov and University of Exeter. Available online at: https://projects.exeter.ac.uk/igov/new-thinking-climate-emergency-declarations-accelerating-decarbonisation/
Howarth C, Lane M, Fankhauser S (2021) What next for local government climate emergency declarations? The gap between rhetoric and action. Clim Chang 167:27. https://doi.org/10.1007/s10584-021-03147-4
ICEF (2021) Governments climate emergency declaration spreadsheet. http://bit.ly/ce-governments Accessed 26 Jul 2021
Kythreotis AP, Howarth C, Mercer TG, Awcock H, Jonas AEG (2021) Re-evaluating the changing geographies of climate activism and the state in the post-climate emergency era in the build-up to COP26. J Br Acad 9(s5):69–93. https://doi.org/10.5871/jba/009s5.069
LaVanguardia (2019) Zeroport exige la paralización de las ampliaciones del puerto y el aeropuerto, 28th November 2019. https://www.lavanguardia.com/vida/20191128/471931605963/zeroport-exige-la-paralizacion-de-las-ampliaciones-del-puerto-y-el-aeropuerto.html
LaVanguardia (2020) ¿Cuál es la contribución de ciudad de Barcelona a la crisis climática? 16th January 2020. https://www.lavanguardia.com/natural/cambio-climatico/20200116/472916021999/barcos-crisis-climatica-emisiones-barcelona.html
Leixnering S, Höllerer M (2021) 'Remaining the same or becoming another?' Adaptive resilience versus transformative urban change. Urban Stud 2021:1–11. https://doi.org/10.1177/0042098021998923
Long J, Rice JL (2019) From sustainable urbanism to climate urbanism. Urban Stud 56(5):992–1008. https://doi.org/10.1177/0042098018770846
MacGregor S (2021) Finding transformative potential in the cracks? The ambiguities of urban environmental activism in a neoliberal city. Soc Mov Stud 20(3):329–345. https://doi.org/10.1080/14742837.2019.1677224
Meo B, Graham S, Ariza E, Casellas A, Delfino D (2021) The resident and visitor gaze: a comparison of coastal social values at risk due to sea-level rise. Environ Sci Pol 123:202–209. https://doi.org/10.1016/j.envsci.2021.05.017
Miller S (2021) Women in power: learning through climate emergency declarations in Canada Miller, Shelley. Master Thesis, Lakehead University. https://knowledgecommons.lakeheadu.ca/handle/2453/4864
Moloney S, Doyon A (2021) The Resilient Melbourne experiment: Analyzing the conditions for transformative urban resilience implementation. Cities 110:103017. https://doi.org/10.1016/j.cities.2020.103017
Nissen S, Cretney R (2021) Retrofitting an emergency approach to the climate crisis: a study of two climate emergency declarations in Aotearoa New Zealand. Environ Plan C: Politics Space. https://doi.org/10.1177/23996544211028901
Patterson J, Schulz K, Vervoort J, van der Hel S, Widerberg O, Adler C, Hurlbert M, Anderton K, Sethi M, Barau A (2017) Exploring the governance and politics of transformations towards sustainability. Environ Innov Soc Transit 24:1–16. https://doi.org/10.1016/j.eist.2016.09.001
Patterson J, Thaler T, Hoffmann M, Hughes S, Oels A, Chu E, Mert A, Huitema D, Burch S, Jordan A (2018) Political feasibility of 1.5°C societal transformations: the role of social justice. Curr Opin Environ Sustain 31:1–9. https://doi.org/10.1016/j.cosust.2017.11.002

Patterson J, Wyborn C, Westman L et al (2021) The political effects of emergency frames in sustainability. Nat Sustain 4:841–850. https://doi.org/10.1038/s41893-021-00749-9
Pelling M (2011) Adaptation to climate change, from resilience to transformation. Routledge, London/New York
Pickerill J (2021) Eco-communities as insurgent climate urbanism: radical urban socio-material transformations. Urban Geogr 42(6):738–743. https://doi.org/10.1080/02723638.2020.1850618
Port de Barcelona (2020) Press Release: El Port de Barcelona lamenta la manipulació que fa l'Ajuntament de Barcelona de les dades d'emissions de CO2 de l'activitat portuària, 15th January 2020. https://contentv5.portdebarcelona.cat/cntmng/gd/d/workspace/SpacesStore/da5b2de6-3eb5-40df-9f6e-0e05583cc882/200115_EmissionsCO2PortdeBarcelona.pdf
PSC (2019) Eleccions municipals 2019: Imaginem Barcelona. https://collboni.barcelona/programa/programa-electoral.pdf
Robin E, Castán-Broto V (2021) Towards a postcolonial perspective on climate urbanism. Int J Urban Reg Res 45(5):869–878. https://doi.org/10.1111/1468-2427.12981
Ruiz-Campillo X, Castán Broto V, Westman L (2021) Motivations and intended outcomes in local governments' declarations of climate emergency. Politics Govern 9(2):17–28. https://doi.org/10.17645/pag.v9i2.3755
Sanz J (2020a) Intervention of Janet Sanz, Deputy Mayor for Environment, Barcelona City Council. Barcelona Climate Emergency Declaration, 15th January 2020, Barcelona City Hall. https://youtu.be/rfW-_zIzVjA (minute 26:00)
Sanz J (2020b) Intervention of Janet Sanz, Barcelona Deputy Mayor for Environment. Press Conference: Sanz i Badia expliquen iniciatives relacionades amb les infraestructures i l'emergència climàtica, 14th January 2020, Barcelona. https://ajuntament.barcelona.cat/premsa/2020/01/13/sanz-i-badia-expliquen-iniciatives-relacionades-amb-les-infraestructures-i-lemergencia-climatica/
Satorras M (2021) From commissioned reports to co-production processes: evolution of urban climate policy-making through the case of Barcelona. In: Leal Filho W, Luetz JM, Ayal D (eds) Handbook of climate change management: research, leadership, transformation. Springer, Cham, pp 4379–4401. https://doi.org/10.1007/978-3-030-57281-5_42
Satorras M, Ruiz-Mallen I, Monterde A, March H (2020) Co-production of urban climate planning: insights from the Barcelona climate plan. Cities 106:1–11. https://doi.org/10.1016/j.cities.2020.102887
Seebauer S, Ortner S, Babcicky P, Thaler T (2019) Bottom-up citizen initiatives as emergent actors in flood risk management: mapping roles, relations and limitations. J Flood Risk Manag 12:e12468. https://doi.org/10.1111/jfr3.12468
Steen T, Brandsen T, Verschuere B (2018) The dark side of co-creation and co-production: seven evils. In: Brandsen T, Steen T, Verschuere B (eds) Co-production and co-creation: engaging citizens in public services. Routledge, New York, pp 284–293
Swyngedouw E (2005) Governance innovation and the citizen: the Janus face of governance-beyond-the-state. Urban Stud 42(11):1991–2006. https://doi.org/10.1080/00420980500279869
Swyngedouw E (2010) Apocalypse forever? Post-political populism and the spectre of climate change. Theory Cult Soc 27(2–3):213–232. https://doi.org/10.1177/0263276409358728
UNFCC (2020) Commitments to Net Zero Double in Less Than a Year (UN CLIMATE PRESS RELEASE, 21 September, 2020). https://unfccc.int/news/commitments-to-net-zero-double-in-less-than-a-year Accessed 15 June 2021
Voorberg WH, Bekkers VJJM, Tummers LG (2015) A systematic review of cocreation and co-production: embarking on the social innovation journey. Public Manag Rev 17(9):1333–1357. https://doi.org/10.1080/14719037.2014.930505
Wilson J, Swyngedouw E (2014) Introduction: seeds of dystopia: post-politics and the return of the political. In: Wilson J, Swyngedouw E (eds) The post-political and its discontents: spaces of depoliticisation, spectres of radical politics. Edinburgh University Press, Edinburgh, pp 1–22
Wolfram M (2016) Conceptualizing urban transformative capacity: a framework for research and policy. Cities 51:121–130. https://doi.org/10.1016/j.cities.2015.11.011

XR (2019) The deliberative rebellion – Film. https://extinctionrebellion.uk/2019/08/27/the-deliberate-rebellion-short-film-with-exclusive-track-from-massive-attack/ Accessed 15 Oct 2021
XR (2021) Extinction Rebellion: About us, Our demands, Our Values. https://rebellion.global/about-us/ Accessed 15 Oct 2021
XR Barcelona (2019a) Letter to the Mayor Ada Colau, 10th July 2019. https://xrbarcelona.org/en/blog/letter-to-the-mayor-ada-colau/
XR Barcelona (2019b) Evaluation of the meeting with Ada Colau, 3rd October 2019. https://xrbarcelona.org/en/blog/evaluation-of-the-meeting-with-ada-colau/
XR Barcelona (2019c) End of year newsletter. https://xrbarcelona.org/en/blog/newsletter-end-of-year-2019/
XR Barcelona (2020a) Climate emergency declaration of Barcelona, 15th January 2020. https://xrbarcelona.org/en/blog/climate-emergency-declaration-of-barcelona/
XR Barcelona (2020b) Valoració i demandes sobre la Declaració d'Emergència Climàtica de l'Ajuntament de Barcelona, 18th February 2020. https://xrbarcelona.org/ca/blog/valoracio-declaracio-emergencia-climatica-ajuntament-barcelona/
Ziervogel G, Pelling M, Cartwright A et al (2017) Inserting rights and justice into urban resilience: a focus on everyday risk. Environ Urban 29(1):123–138. https://doi.org/10.1177/0956247816686905

Chapter 9
Contested Spaces for Negotiated Urban Resilience in Seville

Ángela Lara García, Luis Berraquero-Díaz, and Leandro del Moral Ituarte

Abstract This chapter aims to analyse collective experiences contributing to urban resilience carried out by self-organised civil society groups in Seville (Spain). We use the idea of contested spaces, embedded in the concepts of radical resilience and negotiated resilience, to analyse the case study. We argue that these practices have promoted, by contestation and negotiation, essential green infrastructure for resilience to hydro-climatic risks in the city. Indeed, these communitarian experiences appear in a context marked by a long history of top-down urban planning, in which urban resilience is traditionally understood as a target to be achieved through a robust hydraulic infrastructure system. With the effects of climate change becoming more and more present, bottom-up civil society initiatives have emerged as a form of community resilience in recent decades. We discuss the processes of transformation and hybridisation of these experiences, their conditions of development as well as the limitations of the communitarian approach related to urban resilience.

Keywords Co-management · Community management · Negotiated resilience · Radical resilience · Urban green infrastructure

9.1 Introduction

Social and community movements have been increasingly demanding for urban green spaces due to the progressive social recognition of their socio-ecological values (Campbell et al. 2016). In the context of the climate emergency, this demand is

Á. Lara García (✉) · L. del Moral Ituarte
Department of Human Geography, University of Seville, Seville, Spain
e-mail: anglargar@us.es

L. Berraquero-Díaz
Social Research and Participatory Action Group (GISAP), Pablo de Olavide University, Seville, Spain

I. Ruiz-Mallén et al. (eds.), *Urban Resilience to the Climate Emergency*, The Urban Book Series, https://doi.org/10.1007/978-3-031-07301-4_9

currently addressed in discussions about urban resilience. The purpose of this chapter is to analyse collective urban resilience experiences in the city of Seville (Spain) that, through self-organisation, contestation and civic negotiation processes, have created a large green space system that could be categorised as a network of contested spaces (Low 2009). Despite the unequal power relationship with urban governmental authorities, we argue that these community experiences represent a set of bottom-up initiatives that, since the 1980s, have played a key role in shaping the urban green infrastructures that today are contributing significantly to resilience against hydro-climatic risks in the city. The idea of community contestation, already integrated in the concepts of radical resilience and negotiated resilience, is used to structure the case of Seville, exemplifying a form of co-production of the city.

These community initiatives have gone through different management stages throughout their history, and in this chapter, we differentiate them as follows. We call community management those initiatives managed exclusively by the community in a self-organised and autonomous way. We also differentiate between two types of co-management with the local administration: one called public-community co-management, when referring to a collaboration between the local administration and community movements, and the other called public-private co-management, when collaboration is carried out with private agents (a company or other operators).

The chapter is structured in five sections. The first one is a theoretical section that analyses the main tensions and different logics within urban resilience approaches. Diving into these approaches contributes to the understanding of the prevalent logics in the development of urban flood defence infrastructures in Seville. It also contributes to analyse the logics behind the interventions of social and community movements to reclaim green spaces in the city.

Second, the tragic historical relationship between the city of Seville and hydro-climatic risks (especially floods and heatwaves) is described as the background of current bottom-up experiences of urban resilience. Despite the modification and occupation of areas of floodplains that have experienced frequent and devastating floods in the past, the city currently enjoys a—perhaps false—feeling of security. The current urban robustness to floods results from a costly historical process of infrastructure expansion and constitutes the context for the development of new collective demands and imaginaries regarding climate resilience.

Third, as the core of the work, we present and analyse the initiatives undertaken by self-organised civil society groups. In a context of institutional neglect regarding the provision of public infrastructure and services, and specifically green areas, these initiatives assumed leadership for the socio-ecological transformation of several areas on the outskirts of Seville. The selection of these initiatives is based on the *Booklet of urban resilience community initiatives in Seville and Barcelona: civil society against the effects of climate change* carried out in the context of the RESCITIES Project (Satorras et al. 2020). On this basis, a review is made of the most significant processes of citizen action that have promoted the conservation or creation of green infrastructures in the city. The existing bibliography on the city's contemporary social and environmental history and our transdisciplinary experience in the subject under study have also allowed us to delve into these pioneering experiences of urban greening in the city. For this purpose, in addition to reviewing specific literature and documents, four semi-structured interviews were conducted with representatives of civil society organisations currently engaged in climate action and governance in Seville (between April and November 2020), and another four unstructured interviews were conducted with key actors involved in community processes (between February and March 2021).

Fourth, as the main conclusions, the study shows that historically, large-scale infrastructure works, especially for protection against flooding, have prioritised safeguarding the interests and spaces of the city's urban elites. In contrast, during the last decades, an alternative approach based on the construction of resilience as a bottom-up and nature-based process reveals the capacity of civil society initiatives to build a more resilient city in terms of hydro-climatic risks. As a result of the spatialised and diachronic analysis of these experiences, we discovered that the process of city co-production is expressed, with different rhythms and durations, through three different management situations, i.e. community management, public-community co-management and public-private co-management. These processes are not conflict-free. Community management and public-community co-management schemes prove to be fragile and costly; they require great collective commitment and strong teams of people who are able to facilitate the processes, promote the values and drive the actions needed to construct this new community climate resilience. Concerning public-private co-management, it leads to the disappearance of the conditions necessary for community resilience, such as the community networks themselves or the feeling of collective ownership and belonging.

9.2 Contested Spaces in Cities for Building Urban Resilience to Hydro-climatic Risks

Multiple scientific reviews in recent years have identified tensions around the concept of urban resilience (Brand and Jax 2007; Meerow et al. 2016; Moser et al. 2019; Chelleri and Baravikova 2020). Meerow et al. (2016), in particular, identify six tensions: (1) characterisation of urban system; (2) notion of equilibrium; (3) resilience as a positive, neutral (or negative) concept; (4) pathways to resilience (i.e. persistence, transitional or transformative); (5) understandings of adaptation; and (6) timescale of action. The tension referring to the pathways of resilience is perhaps the oldest, although it is still relevant. The academic literature on resilience has used two metaphors to show this tension, i.e. the bouncing-back metaphor and the bouncing-forward metaphor, which have been very well explained by Chelleri and Baravikova (2020). The bouncing-back metaphor, developed mostly in engineering and classical ecology, represents a perspective on resilience that places greater emphasis on system robustness and the ability to absorb perturbations before system change occurs. But it also shows a way of dealing with risk. This approach prefers a risk control strategy that allows a return to the previous state as soon as possible and with little change to the system. On the other hand, the bouncing-forward metaphor represents an approach that includes two possible paths: transition and transformation. This approach addresses risk through management strategies, which incorporate elements such as uncertainty and the ability to adapt and/or transform to different states. In other words, it considers state change as a possible or feasible option. The two metaphors are opposing but coexist today, especially in terms of public policy, as we will illustrate with the logic of contemporary public administrations in Seville when implementing infrastructures to adapt to hydro-climatic risks.

The logic of transformation also helps us show a question that is not yet well developed in the literature on urban resilience but has always been obvious to urban social movements. Cities result from a conflictive process involving actors with capacity to influence from the top-down and also actors, such as social movements, with the capacity to *build and transform the city* from the bottom-up and, therefore, to influence its resilience (Davies 2013; De Carli 2016; Delgado Ruiz 2017).

Traditionally, urban planning and public policies developed in many cities have incorporated an engineering perspective on resilience focused on robustness (Abbar et al. 2016). This perspective has put the focus on the infrastructural, technical and environmental urban challenges exacerbated by climate change rather than on social, equity and conflict-related challenges. In fact, as we will see below, most of the urban infrastructures developed in Seville by the public administration to cope with hydro-climatic risks reproduce the robustness approach. It has allowed urban planners and policymakers to plan the city without considering the problems that urban development itself generated nor analysing the socioeconomic consequences that the very act of *defence* against risk generates on the city and the people who live there. When many of these infrastructures to prevent flooding of the main river and its tributaries were developed during the twentieth century in Seville, the concept of resilience was not as present as today among decision-makers. However, we can argue that they followed the logic of bouncing-back. Only recently, the local public administration in Seville has incorporated strategies for risk management, transition or transformation towards more sustainable and equitable use of public space.

In addition to the six tensions identified before, we must add the tensions recognised from a political perspective, which problematises who decides who should become more resilient, in the face of what, how, when, where and at whose expense (MacKinnon and Derickson 2012; Chandler 2014; Vale 2014; Meerow and Newell 2016; Kaika 2017; see Chap. 5 for the results of this the analysis in the Latin American context). The political perspective in urban resilience focuses on understanding the power dynamics that have often been overlooked in the literature and in the practical application of this approach. However, the tensions between who has the power to decide about resilience indicate the difficulties in operationalising the concept of resilience (Davoudi et al. 2012). The use of resilience as a synonym for urban sustainability further complicates its application, with the danger of becoming an unusable analytical framework, or only usable as a metaphor or a bridging concept, rather than as a framework that can generate specific applications for urban policies (Pearson et al. 2014).

To make the debate even more complex, there is a growing trend to inherently attribute to resilience a positive connotation (Meerow and Stults 2016). Resilience is a characteristic of systems, but it does not provide information about the desirability of that system. For example, some urban infrastructures developed in Seville in response to river flooding have systematically forgotten certain neighbourhoods with lower incomes (such as the Triana neighbourhood), thus aggravating their socioeconomic vulnerabilities as well as their hydro-climatic vulnerabilities. Similarly, some highly resilient urban socioeconomic and cultural dynamics are not necessarily desirable for all those who have to endure them and may even compromise the future sustainability of the ecosystems on which these people depend (Dornelles et al. 2020). Thus, the harsh criticism that many authors have made of resilience considers that it consolidates a conservative and neoliberal urban agenda while neglecting issues related to power relations, social inequality and distributive justice in the face of risks and threats (Bahadur and Tanner 2014; Kaika 2017).

This is why the literature on the so-called community resilience insists that other approaches to urban resilience must be included (Mulligan et al. 2016). Developing a public policy or carrying out an intervention using the urban resilience approach can have impacts on different population groups at different temporal and spatial scales. The community resilience framework asks to identify these impacts and how these policies should incorporate the local and/or regional communities affected (Berkes and Ross 2013). It also questions how a community can best cope with specific risks. Further, there are not only one-off events or shocks, such as earthquakes, floods or fires, on which to build resilience for recovery, transition or transformation but also chronic threats, such as unemployment, lack of access to affordable housing or poverty, that weaken communities and thus their resilience (Koliou et al. 2020). The community resilience framework highlights that we must overcome the idea of a return to a previous state when talking about urban resilience, taking into account that there are situations, especially in the urban context, where it would mean bouncing-back to social inequity and unsustainable or undesirable socioeconomic systems (Fazey et al. 2018). Indeed, the inclusion of equity and social justice criteria when applying the urban resilience approach to urban infrastructure and public policies is still an issue to be addressed. We show below some recent attempts to address this gap in the academic literature.

Some authors use the concept of socio-ecological vulnerability to show how climate risks and hazards affect communities living in cities differently, to ensure that the solutions proposed for climate mitigation and adaptation include the needs of these diverse communities (Anguelovski et al. 2016; Shokry et al. 2020). From this perspective, vulnerability to climate risks stems from a combination of social vulnerability defined by sociodemographic variables (e.g. poverty, unemployment, age, minorities, etc.) and ecological vulnerability defined by biophysical environmental factors (e.g. exposure, infrastructure characteristics, etc.) (Lara Garcia et al. 2020).

Other authors analyse urban resilience trade-offs produced due to social, environmental, public policy and urban infrastructure interventions that have supposedly been implemented to improve the adaptive capacity, socio-ecological sustainability and/or social justice of a system (Chelleri et al. 2015). These trade-offs can occur at different spatial scales and affect different groups in different ways or even across time scales. However, the trade-off framework does not consider the intersectional inequalities within a society, which can be based on sex, gender, ethnicity, class or age. Nor does it consider the role of historical injustices in shaping a community's system of relationships and identities. These shortcomings can determine individual or collective vulnerability in a crisis or the ability to access resources and even the capacity of some groups to participate in decision-making. In fact, the trade-off framework does not address procedural equity in participation; that is, it does not ensure that urban resilience plans meet effective and inclusive participation agendas with a real possibility of having some influence on decision-making (Meerow et al. 2019).

Other works related to equity and social justice, such as that by Harris et al. (2017), highlight resilience's procedural dimension through developing the concept of *negotiated resilience*. From this perspective, urban resilience is understood as a process that includes actors with multiple and sometimes opposing interests and multiple notions of what should or should not be resilient in an urban setting. This multiplicity of actors means that resilience is not understood as an objective per se, but rather as a process, a broad negotiation process in which the participation and inclusion of all the people affected by urban resilience decision-making receive explicit support. However, this framework acknowledges that the unequal correlation of forces among social actors does not guarantee equity in the negotiation process (Ziervogel et al. 2017).

Moreover, there are contexts in which institutional neglect results from a supervening event, such as after natural disasters. There are also contexts in which this neglect is the cause of a deliberate lack of institutional interest, e.g. in some areas of the urban periphery, as we will discuss later in this chapter. For all these reasons, how urban conflict is managed will be key to incorporate justice and equity in the framework of resilience. In cities, many people live together with different processes of collective identification of the urban space itself, and, therefore, conflict is an inherent part of the processes of urban co-production (Delgado 2004). It is also the result of control or hegemony over the uses of the urban space, which are affected by global and local market logics (Gramsci 1999; Davies 2013). Starting from the idea that there are different ways of conceptualising the urban, we understand that there must also be a variety of ways of analysing and managing conflict. Capel (2003) recovered the Roman tradition in the denomination of these urban dimensions by affirming that cities are at the same time *urbs*, *polis* and *civitas*. The *urbs* is the built space, what today we would understand as urban infrastructure and urban metabolic flows. The *polis* would correspond to the formal and informal government structure and the political-administrative units that influence urban management. But the city is also the result of the different practices carried out in it by the people who inhabit it, the *civitas*. A city does not exist without its inhabitants and their experiences in urban spaces. That is the reason why it is important to highlight the role that citizens and particularly urban social movements have played and continue to play in transforming the city (Capel 2003). To elaborate on this idea,

Davoudi et al. (2012, p.309) suggest that resilience has *the potential to develop as a more radical and transformative agenda* by recognising the role of bottom-up initiatives, such as those set in motion by social and community movements. To delve deeper into this idea, Jon and Purcell (2018) have proposed the notion of *radical resilience* using the theories of agonistic and anarchist planning. These authors believe that agonistic conflict between government and citizens ignites a spark for community management, i.e. when a community claims a space as its own, and the municipality tries to appropriate the management of that urban space, as we will see later in this chapter. These authors consider radical resilience the process of empowerment in which citizens realise that they can *self-manage their own affairs* (Jon and Purcell 2018, p.1).

As seen in this section, the literature on urban resilience has focused more on the *urbs* and the *polis* following Capel's categorisation. Some authors have incorporated the consequences on the *civitas* of interventions on the *urbs* or the *polis*. However, only some of the above-referenced studies have focused on the city-building power of citizens, who live in the city and contest it, giving a new meaning to urban spaces supposedly created for a specific purpose, to subvert them and re-appropriate them. In what follows, we focus on the spatial practices and representation of urban spaces set in motion when the neighbourhoods of Seville's urban periphery were configured since the early twentieth century, considering this process as a combat for urban resilience. We argue that this process of contestation could be framed in what anthropologist Setha Low categorised as *contested spaces* (Low 2009), since the conflict (i.e. the process of contestation) did not take place in urban space, but *for* urban space. As we will show, social action aimed at demanding and projecting a *way of being and experiencing the city* (Lyons et al. 1980; García Jerez 2011; Huang 2019). Thus, with their subjective and collective urban practices and experiences, and even through occupation and self-organisation practices, the communities of the peripheral neighbourhoods demanded and created not only urban green infrastructure but also other types of basic community services that were later consolidated in the city. Properly designing and maintaining this *social infrastructure* (Klinenberg 2018) might be a powerful strategy for a more equal and resilient society. In other words, the processes of social contestation played a fundamental role in shaping Seville's urban green and social infrastructure, which has ultimately become a key factor in addressing the hydro-climatic risks that the city is currently facing. Moreover, these contestation processes followed logics of transformative resilience that contrast with those of bouncing-back resilience corresponding to the interventions developed by the public administration throughout a previous trajectory and even at the same time. We argue that long before the city's commitment to resilience was on the public agenda of Seville, and even before it was explicitly incorporated into activist and community initiatives, the mobilised citizens organised to transform the urban space, which in turn reinforced community resilience.

9.3 Seville and Hydro-climatic Risks: The Construction of a Robust and Self-Confident City

Very intense socio-natural risks and opportunities for human wellbeing are combined and concentrated in the city of Seville due to its situation, location and territorial resources and the nature and history of its urban functions. The city is situated at the end of the Guadalquivir River basin that, with a surface area of 57,000 km^2, spreads over a valley of fertile land. The river catchment is exposed to ocean winds and enjoys mild winters and relatively abundant, although recently decreasing, rainfall (around 600 mm/year). The city is located 80 km inland, at the last fordable point of a navigable estuary, the only river port on the Iberian Peninsula, and at a crossroads for terrestrial communication. These conditions explain the existence and importance of the city as the economic capital of the Hispanic Empire (sixteenth and seventeenth centuries). Even today, Seville is the principal city in the south of the Iberian Peninsula.

The Guadalquivir River is more like a gigantic creek than a river (Vanney 1970; Del Moral 1991), as the relatively abundant winter rain is characterised by its great seasonal and inter-annual irregularity. Before the infrastructural transformation carried out throughout the twentieth century, the river instantaneous flow fluctuated between 10,000 and 10 m^3/s, with a yearly cumulative flow of between 20,000 and 500 hm^3/year. Most of Seville's urban space, protected by city walls since antiquity, is less than 7 metres above sea level, in an area where river floods higher than this height used to occur every 10 years. Apart from the flooding caused by the river overflowing, Seville has also had to contend with stormwater flooding (*internal flooding*). Whenever the water outside the protected city rose to a height above the ground level inside the walls, it was not possible to discharge the rainwater through the traditional drainage network. On the contrary, the external floodwaters entered the city through the drainage system, aggravating water accumulation in the lowest areas. This, together with its long history as an urban centre, explains why Seville has one of the longest and most detailed chronicles of documented flooding in the world (Guichot 1877; Palomo 1878; González Dorado 1975; Solís 2022).

To complete this general characterisation, it should be noted that mild, frost-free winters are combined with torrid summers (temperatures of 45 °C are not unusual on some days in July and August). The mean annual temperature in the city (Fig. 9.1) shows an upward evolution throughout the period 1940–2016 (statistically significant at $p < 0.01$). After 1994, the mean temperature has always exceeded 16.0 °C. By contrast, throughout the period 1940–1994 (40 years out of a total of 53), it was frequent to find lower temperatures (mean temperatures below 16.0 °C). Comparing the two halves of the series, the mean annual temperature has gone from 15.6 °C in the period 1940–1980 to 16.1 °C later (CHG 2021). Regarding heatwaves, the dates show an increase in frequency and a prolongation in the number of heatwave days, as well as an advance in the months of appearance (May–June) before the hottest period (July–August). Concern about the consequences for the population's health, especially for the most vulnerable sectors, is growing (Junta de Andalucía 2019).

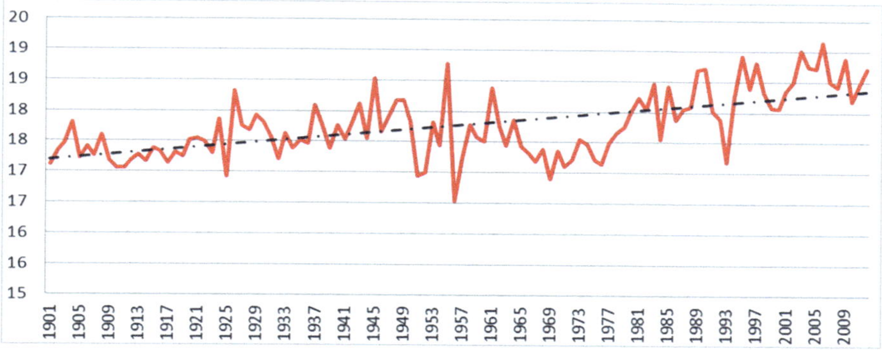

Fig. 9.1 Mean annual temperature in Seville (1901–2010). (Source: author's elaboration based on data from CRU TS 3.21)

Given the content of the case study developed in the following section (community-based initiatives of urban green infrastructure relying on community management and public-community co-management), in this background and context synthesis, we do not address the issues related to drought, which also seriously threatens the city. Since the spatial scale at which drought is generated surpasses the urban scale, as well as its impacts and the responses it provokes, drought is not a central issue to the experiences exposed as flooding and heatwaves.

9.3.1 *The Defence Against Floods*

Although earlier preventive actions were developed, it was not until the beginning of the twentieth century that the execution of the public works started to radically transform the hydrographic network of Seville. This network includes both the channel of the main river, the Guadalquivir, and its torrential tributaries on the left bank (i.e. Tagarete, Tamarguillo, Guadaira). The systematic diversion, channelling or tubing of these streams was accompanied by the construction of a belt of dikes with a height of 12 metres above sea level that extends the defended area from the 260 ha of the historic urban site to the current 5280 ha (Fig. 9.2).

The major stages of the construction of this hydraulic infrastructure are the so-called Moliní and Sanz Larumbe Plan (1903–1929), the Delgado Brackenbury Plan (1929–1950), the diversions of the Tagarete and Tamarguillo (1930 and 1962), the channelling of the Guadaira (1977) and the Cartuja Channel (1975–1982). This was a long and costly process interspersed with catastrophic floods, such as those in 1917, 1926, 1945, 1952 and 1963, and many others less traumatic. Repeated catastrophes revealed the fragility of the system and the increasement of risk, in terms of people affected, throughout this period. Despite the hydraulic infrastructures, the enormous urban expansion during the twentieth century took place mostly over flood-prone lands. To a certain extent, this risk was addressed by the accumulation

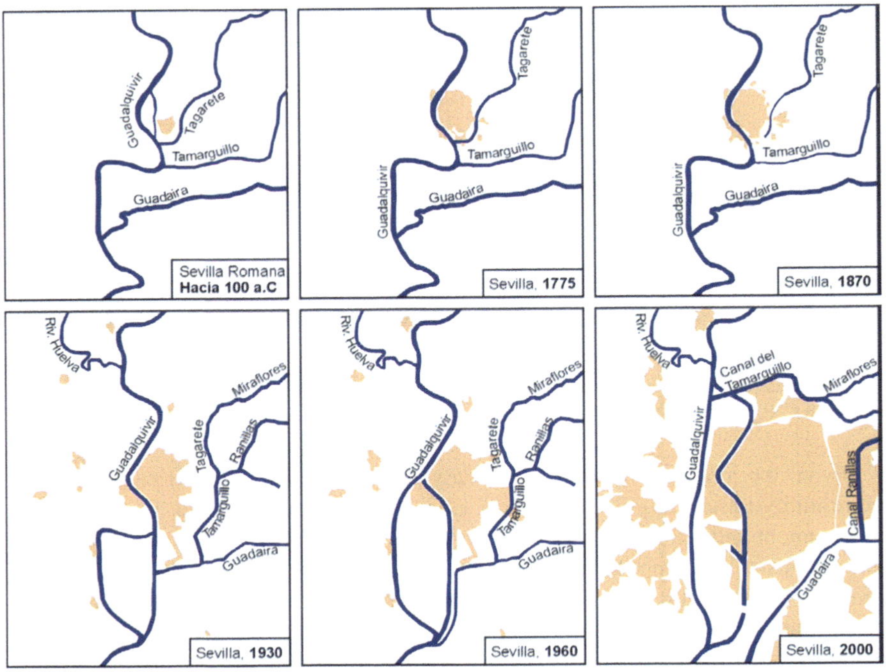

Fig. 9.2 Guadalquivir River with its tributaries as it passes through Seville with the modifications made to their courses. (Source: García García (2004))

of technical-administrative initiatives (action protocols, personnel and resources allocated ad hoc) and local knowledge (consolidated practices of solidarity and mutual support) that allows the implementation of actions to reduce impacts both during and in the wake of the catastrophes themselves (Solis 2022). However, the catastrophes served above all to justify the continuous development of the hydraulic infrastructure programme led by the government authorities as a condition for consolidating the urban expansion of the city without taking into account the real physical potential of the affected hydrographic network (García García 2004).

Throughout that process of infrastructure development, there has been an intense citizen debate about the conception and technical design of the different alternatives, their objectives and priorities, privileged and neglected areas as well as their costs and benefits. In this regard, it is meaningful to highlight that once the Seville Port was expanded and modernised (Moliní Plan 1929), the popular neighbourhood of Triana continued to be exposed to flooding for decades (until the Delgado Brackenbury 1950 Plan). In the same sense, the channelling and diversion towards the south-east of the streams that crossed the city (i.e. Tagarete and Tamarguillo), which enabled the urbanisation of a large area for the 1929 Ibero-American Exhibition, increased the vulnerability of the working-class settlements that had arisen on the east bank of these streams (Díaz del Olmo et al. 2012). Both of them are examples of the hegemony of certain urban-territorial approaches with sometimes internal conflicting interests, i.e. urban expansion and Port consolidation. Nonetheless, underneath these discrepancies lies a common hard infrastructural strategy for fighting the *atrocious, ferocious and destructive* swells of the Guadalquivir and its tributaries and other similar expressions used to refer to them in the discourses used by government authorities. At the same time, a big part of the population had to suffer the vast and unevenly distributed impacts of this mode of coping with the *natural threat*. In the decades of the 1940s, 1950s and 1960s, fast and unordered urban expansion over flood-prone areas, along with unfinished or poorly designed hydraulic and urban infrastructure, produced large masses of refugees to the urban periphery (the so-called Seville of the shelters), in a post-civil war, dictatorship context (Solís 2022).

Even today, new projects to expand the flood defence system exist, such as the rechannelling of the Tamarguillo in the north or the new lock of the Port, built in 2010 in the south. However, since the last flood in 1963, the city has not been affected by any serious floods, apart from small marginal (in spatial and social terms) areas outside the protected enclosure and apart from water ponding problems caused by heavy rains (see next section). Along with the local defence against flooding, the Guadalquivir River Basin Authority has carried out large-scale hydraulic works to regulate the river flow, building a reservoir system of more than 40 big dams, with a total capacity (8500 hm^3) greater than the average annual flow of the river (6000 hm^3/year). In addition, the course of the estuary downriver from Seville has been straightened (with the river course shortened by 20 km), which facilitates the drainage of river floods. The main drivers behind the former were agricultural interests (e.g. irrigation), while behind the latter were Port interests. These two agents—irrigation and navigation—along with urban development promoters have been aligned in their strategy to control nature, albeit not without contradictions and partial conflicts between them.

9.3.2 Internal Flooding and Rainwater Management

Internal flooding has always been another risk in the city's history and continues to be so today. It originates from the inability to discharge stormwater, especially when Seville was surrounded by floodwaters that entered the city through the ramshackle traditional drainage system. The slow modernisation of this system gradually reduced the problem in the city's central areas. It was not until January 1976 that the municipal water supply and treatment company (EMASESA) passed a comprehensive water drainage plan for Seville to be developed during the 1976–1982 period. The plan's execution was prolonged until 1989 and included the construction of 23 large drains or discharge pipes, pumping stations and 4 wastewater treatment plants (Del Moral 1991; Solís 2022). Even so, nowadays, waterlogging problems persist in some city areas. This, together with new factors that affect the quality of wastewater, has recently led to new hard infrastructure developments, such as rainwater retention tanks, that clash with the discourse favouring *nature-based solutions*, such as sustainable urban drainage systems (Lara 2018; EMASESA 2020).

9.3.3 Heatwaves

Another major hydro-climatic risk in Seville is the heat and, more specifically, the periods of extreme heat, heatwaves, in the context of global warming. The torrid temperatures in the Guadalquivir valley during long summers that can last 5 months are an intrinsic characteristic of the city. While historically valued and experienced as an inconvenience, it has never been perceived as a *catastrophe* as severe as floods. Sevillians have historically developed a well-known and socially valued traditional culture of adaptation to heatwaves that includes adjustment of urban and building typologies to heat, adequation of timetables and types of activities and seasonal changes of residence for those able to afford them. Still, there is unequal access to these adaptation strategies (Figueroa and Suárez-Inclán 2008).

Social acceptability and resilience towards periods of extreme heat have decreased over recent decades, even before the current increase in the frequency, duration, temperature levels and temporal location of the heatwaves attributable to climate change. Since the end of the twentieth century, and especially since the 1990s, the currently dominant infrastructure and technology-driven response to heat, i.e. air-conditioning, has become very widespread in the city. The use of air-conditioning today, unlike decades ago, is considered socially essential in any public space, work centre and means of transport and by a very high percentage of private (not necessarily affluent) households. Public interventions have been fundamentally directed in this direction.

A significant and intense social debate is taking place around new focuses, diagnoses and alternatives to the dominant technology response for facing heatwaves. For instance, in an interview published in the local media in Spain, Prof. Enrique Figueroa leading the University of Seville's Sustainability Unit calls for the recovery of what he names *good traditional practices*: *We must carry out a study of pedestrian areas to implement measures to make them more pleasant and introduce climatic criteria in development plans. We must analyse which are the best materials and create shadows with suitable trees, pergolas or other structures. We must make the city more liveable, without forgetting the most disadvantaged neighbourhoods and outskirts that have enormous deficiencies* (Parejo 2021, p.2). In a return to adaptive local knowledge and the culture of co-existing with the heat (e.g. shade, vegetation, building orientation, incorporating water into the home, etc.), some collective experiences and projects in the city are working to put into practice these imaginaries and underpin debates on how new community strategies generate resilience. These experiences are presented in the following section.

9.4 Resilience as a Transformative Process: Social Contestation Around Seville's Urban Green Infrastructure

Since the time of ancient civilisations, vegetation has been a common component of the urban system (Guerrero et al. 2016). Mediterranean cities have traditionally included garden areas that combine elements of vegetation with flowing water, with an obvious social and climatic function, while shaping their cultural identity.

Discourses supporting the greening of urban spaces and nature-based solutions are increasingly gaining prominence in urban resilience policies, especially as adaptation strategies to hydro-climatic risks (WHO 2016; Gutiérrez et al. 2017). The search for the benefits provided by nature has also led to the evolution of the concept of urban green spaces: from the previous perspective of gardens as a means of beautifying public space to the current vision that seeks to generate infrastructures capable of providing urban ecosystem services. Thus, urban green infrastructure is defined as a *system of natural spaces and processes to support urban resilience and the quality of life* (Juvillà 2019, p.20). This definition helps to connect and give coherence to the elements that compose the natural heritage of urban areas: from the street trees to natural habitats on the sub-regional scale.

Urban parks emerge as core components of this green urban infrastructure from the point of view of constructing resilience against hydro-climatic events (Gutiérrez et al. 2017; Juvillà 2019). The large volume of plants and water in urban parks helps to create shade and absorb heat through evapotranspiration.[1] When properly irrigated, urban parks are able to create a microclimate that mitigates the heat island effect occurring in cities (Guerrero et al. 2016). Water retention is also a key aspect, especially in Mediterranean climates, where vegetation maintains its evapotranspiration ability as long as it is able to extract water from the soil. From this perspective, urban parks are large areas of permeable land that provide a territory with an improved capacity for infiltration, thus contributing to water retention and aquifer recharge. Adequate integration of green infrastructure with the urban water network helps reduce exposure to flood risk and retain water resources to avoid drought.

The contribution of parks to urban resilience is underpinned by their attributed social value as places to connect with nature. They are places where recreational, cultural and educational activities can take place in the open air, thus facilitating social relationships and connection with the environment (Gutiérrez et al. 2017). An example of this has been the large number of people visiting parks during the COVID-19 pandemic (Kleinschroth and Kowarik 2020). In the context of mobility restrictions caused by the pandemic, these spaces enabled the population to find the connection with nature that they needed during local lockdowns (Fig. 9.3).

[1] The transpiration process of vegetation can dissipate between 30 and 70% of net solar radiation falling on the surface during the hottest period, turning it into latent heat of vaporisation and bringing down the temperature by up to 10 °C on hot, dry days (ACMG n.d.).

Fig. 9.3 Large affluence of people to Alamillo Park (Seville) during COVID-19 derived local lockdown. (Source: Ángela Lara (27 March 2021))

In what follows, we show the urban dynamics involved in the construction of Seville's current green infrastructure (*urbs*), by focusing on the institutional initiatives (*polis*) and social processes (*civitas*) as an example of urban co-production. Firstly, the main milestones and public actions for constructing the urban green infrastructure are outlined. Secondly, the social mobilisation that motivated these public actions is described, as well as the evolution of the co-production processes linked to them.

9.4.1 Conception and Construction of the Seville City Green System

The presence of green spaces has been part of Seville's city landscape from the very beginning of the city history. Examples of these early interventions are the gardens of the Alcázar and the *Casa de la Contratación* palaces and the *Alameda de Hércules* (1574), which is said to be the oldest public garden in Spain and even in Europe.

The first large park in the city, Maria Luisa Park, was built in the first decades of the twentieth century. It was part of an operation to expand the city to the south through the large bourgeois development constructed for the 1929 Ibero-American Exhibition. This intervention was in line with the above-mentioned project to expand the Port to the South (Molini Plan 1903). Later on, the city's first Urban Plan in 1946 (the Bidagor Plan) set down the conceptual bases of the city's current green space. Based on zoning theories, this plan sought to make the municipal territory functional by surrounding building space with a *green belt* that, like a wedge, sliced into the consolidated residential area (Martín 1996). Despite not having much impact, the 1946 Plan proposed an urban model that would be rolled out in the following general plans (1963, 1987, 2006). The 1963 General Urban Management Plan (*Plan General de Ordenación Urbana—PGOU*) subsequently finalised and formalised the layout of the green space system. The city's main parks were sited in some important areas on the city's periphery, many on the courses of former rivers and streams from which the parks would later take their names (i.e. Miraflores, Tamarguillo, Guadaira; see Fig. 9.4).

Fig. 9.4 Seville City Green System. 1. Miraflores Park; 2. Tamarguillo Park; 3. Infanta Elena Park; 4. Amate Park; 5. Guadaira Park; 6. Maria Luisa Park; 7. Buhaira Gardens; 8. Alcázares Garden; 9. Los Príncipes Park; 10. Vega de Triana Park; 11. American Garden; 12. Alamillo Park; 13. San Jerónimo Park. (Source: author's elaboration)

Although the city green space system was conceived by planning in the 1960s, until the 1980s, the City Council had only developed green spaces in the bourgeois areas of urban expansion where the more privileged classes lived (i.e. the parks of Maria Luisa and Los Príncipes). Meanwhile, in the working-class periphery of the city, the large housing estates lacked basic facilities, which became the main demands of neighbourhood associations. Thanks to pressure from the powerful community movements, in the 1980s, work began preparing the land for the Amate and Miraflores Parks (Díaz 2010). Initially conceived as low-maintenance green areas, there was no overall project to link them to the historical and socio-cultural characteristics of the areas, as demanded by community movements (CPPEM 1992).

The large investments in public works due to the 1992 Universal Exhibition solved some of the issues arising from rapid urban expansion (Díaz 2010), including the conversion of the Guadalquivir riverbed as it passes through the city into a basin through the construction of the Cartuja canal. This action enabled the riverbanks to be turned into long walkways, and two new parks were built in areas near the Exhibition site (San Jerónimo and Alamillo). In addition, the historical Buhaira Gardens were restored in the heart of the Nervión district, next to the city's new financial area (see Fig. 9.4).

However, these significant investments did not have any impact on the large brownfield sites on the urban periphery. It was not until the early 2000s that, due to community mobilisation, the City Council used European funds to develop new green areas in the peripheral neighbourhoods (Constenla 1999). A large part of these funds was invested through the signing, in 2003, of an agreement between the Seville City Council and the Guadalquivir Basin Authority to implement a plan for *Hydrological and Forest Restoration and the Protection of Water Courses in the Municipality of Seville.* The objective was to restore 425 hectares of green areas linked to the city's hydrographic network. The plan's actions duplicated the city's green areas between 2004 and 2014 (Jiménez 2014). Nonetheless, the projects provoked several conflicts between citizens and the City Council. For instance, in the context of the 2008 financial crisis, some of these parks were opened without the necessary equipment (e.g. lighting, fountains, benches, etc.), which generated new social protests and mobilisations (Jiménez 2014).

The investments from this Hydrological and Forest Restoration Plan improved some water ecosystems and their integration into the city. However, the endeavour to integrate the green space system into the watercourse network was limited, although the 2006 Urban Plan recognised this problem and included the need for a Green Belt, as already established decades before in the 1946 Bidagor Plan.

Social movements are currently calling for the Green Belt project. They see the project as a strategic opportunity to demand participation in the rehabilitation and management of urban stretches of rivers, which would provide a green infrastructure capable of contributing to urban hydro-climatic resilience. On the other hand, Seville's grassroots climate movement *Movimiento de Entidades por el Clima de Sevilla* has presented a proposal to demand real citizen participation in the co-design of the Green Belt. They have also demanded a commitment from the City Council that this green belt will be ecologically and functionally connected to the rest of Seville and the green system of its metropolitan area (Ameneiro 2021).

9.4.2 From Demands to Management: Community Resilience Through Community Management and Public-Community Co-management of Urban Green Spaces

In Seville, the process of creating green spaces has become a common goal and a key component of social transformation and civic empowerment, mainly in the traditional working-class neighbourhoods. Civic recognition of their social and environmental values has boosted the historical claim for green spaces. It has also given rise to collective urban experiences that have resulted in a green infrastructure system in the form of *contested spaces* (Low 2009). These spaces have become crucial in shaping the current configuration of the city.

Table 9.1 Community-led initiatives in Seville committed to the creation of green infrastructures. CM and PCCoM mean community management and public-community co-management, respectively, while PPCoM refers to public-private co-management

Citizen initiatives	Green space typology	Mobilisation start date	Construction date	CM or PCCoM period	Current situation
Valle Gardens	Historic garden	1977	Fifteenth century	–	Government management
Miraflores Park Committee	Urban garden and park	1983	1991/1999/2011	1991–2016	PPCoM
Tamarguillo Park Committee	Urban garden and park	1990	1999/2007	2006–2012	PPCoM
San Jerónimo Gardens	Urban garden and park	–	1992	1996–present	CM (gardens)
Green and Public Tablada	Floodplain—metropolitan park	2003	–	–	Private property
Rey Moro Garden	Urban garden	2004	2004–present (self-construction)	2004–present	CM
Living Guadaira Park Platform	Park	2002	2014	–	PPCoM
Green Torreblanca Civic Project	Public spaces naturalisation	2018	2018–present (self-construction)	2018–present	PCCoM

Table 9.1 describes some examples of such citizen involvement in the defence and creation of the city's green spaces from the 1980s up to the present day. The leadership of the local community and the capacity of the processes to become transformative tools that improve people's living conditions while coping with climate stressors in these neighbourhoods make these experiences clear examples of community climate resilience and urban co-production processes.

As we will see below, urban gardens in Seville are closely linked to and explained by the active role of community movements in the construction of large parks. In the recent history of the city, urban gardens have been understood as catalysts that spark the emergence of new social subjects (*civitas*) that are able to generate production and social management dynamics in the urban habitat (*polis*), with some specific proposals for an alternative city model (*urbs*) (Dimuro et al. 2013). To show this idea, we focus on one of the most significant examples of this type of practice: the process of co-production and co-management of the Miraflores Park. We discuss this case in relation with other similar experiences, in which social mobilisation has also promoted processes of community urban resilience linked to the construction of the urban green infrastructure in Seville. Throughout such discussion, we identify three characteristics that are common to these processes:

- Cross-cutting nature combining aspects of environmental, urban, social, heritage and collective identity construction
- Social significance and willingness to spread knowledge of historical and natural tangible and intangible heritage, from traditional gardens to river courses, including old buildings, water infrastructure and archaeological artefacts
- Marked proactive and, at the same time, rebellious character, reflected in the launch of provocative agroecological urban garden projects, which also enable citizens' active participation in the design and construction of the city

As mentioned above, one of these processes, undertaken by the Miraflores Educational Park Committee (CPPEM, *Comité Pro-Parque Educativo Miraflores*), has generated major interest due to its long history of over 30 years and its significance and projection on the local, national and international levels.[2] This experience is a prominent example of the construction of climate resilience as a result of a process of social contestation, combining proposals for *community, radical* and *negotiated resilience* having a great influence and major repercussions on the city's urban and social dynamics.

The CPPEM was formed in 1983 by different neighbourhood associations self-organised to promote the Miraflores Park creation through civic participation (Pozo 2019). This community organisation, where women played a fundamental role, demanded the park as a necessary green space in a newly constructed, densely populated, poorly skilled neighbourhood. The park was understood as an instrument of empowerment and social transformation and a possible solution to the social

[2] It was presented at the United Nations Habitat II International Conference held in Istanbul (1996).

problems that existed. Moreover, it aimed to recognise and enhance the natural and cultural heritage of this part of the city periphery, linked to an agricultural past and on which a cultural neighbourhood identity could be built. In this way, the methodological proposal for the construction of the Miraflores Park explicitly stated the need to transform the traditional relationships between citizens, politicians and technicians, turning the neighbours in the area into active participants in the park's creation (CPPEM 1992) (see Fig. 9.5).

In 1991, based on a great capacity for collective self-organisation and in the absence of institutional initiative, the CPPEM took direct action by occupying the *Finca de Miraflores* (named *Casa de las Moreras* and almost 3 hectares of land), where they began the first garden-school projects. The strong social support and great success of the initiative forced the municipal government to formalise the process in an agreement in 1994 to cede the use of land in the area to the CPPEM and provide them with funding for the garden-school project. The involvement of a contracted technical team and the community movement consolidated this project. Together with the space generated around the *Casa de las Moreras*, it became a focus for environmental awareness and an important social engine for action in the neighbourhood. In addition, this initiative developed in Miraflores became a reference for other groups and cities promoting climate resilience strategies.

Following the example of the CPPEM, a second project for social gardens in the *Parque de San Jerónimo* was initiated in 1996. A grassroots socio-environmental

Fig. 9.5 Community tree planting in Miraflores (1993). (Source: Isabel Muñoz)

collective, *Ecologistas en Acción*, proposed the project and signed a similar agreement with the City Council. These two projects were consolidated and grew in the following years. In 2004, a progressive coalition entered the municipal government and launched a participatory budget scheme, which led to the development of three new social gardens (*Parque del Tamarguillo*, *Torreblanca* and *Jardín del Rey Moro*).

Thus, the social gardens became part of a community strategy to demand the conversion of the city's urban wastelands into public parks, preceded in some cases by the occupation of the areas by local residents. During the brief experience of participatory budgeting in Seville (2004–2010), municipal funding of these projects allowed organisations to act with a certain degree of independence, handing over responsibilities to the community organisations that provided the gardens. Citizens designed and built the projects with the understanding that the gardens were a tool for the collective dynamisation and construction of green spaces (Interview 1).

As the projects became established, the demand for green areas increased, and some new community initiatives emerged, such as Guadaira Park and Vega de Triana. Subsequently, in 2009, the Urban Garden Platform (*Plataforma de Huertos Urbanos*) was created as a coordination instrument, becoming a key stakeholder in promoting projects contributing to climate resilience. Despite notable differences in the urban context and organisation model of the collectives, the Platform had a common goal: *to link the neighbourhood with the will to build reality and city around projects with a great potential for transformation* (Interview 1). Based on its experiences, the Platform detailed its proposals in a Master Plan for Gardens (2009), which was presented and discussed with municipal officials. However, it was not formalised in any binding agreement with the City Council (Interview 1).

The lack of any institutional commitment to formalise the projects in the long term, the 2008 economic crisis and the new conservative local government elected in 2011 brought a radical change in the relationship with the City Council that stopped funding the projects (except the Miraflores project that suffered a significant reduction in its funding). In the case of Tamarguillo social gardens, a high level of confrontation was developed against the municipal government, bringing power imbalances into play. The experience was halted by the City Council in 2012 by deposing the team that facilitated the process (Ameneiro 2012). Other projects like San Jerónimo or Rey Moro social gardens continued through community management. So, from 2012 onwards, not only the garden models started to diversify but also their survival strategies.

As mentioned above, the Miraflores Park had an almost 30-year history of community mobilisation behind it and 20 years of managing garden programmes. It had progressed from the initial demand and creation phases to a management phase (Carmona 2004). Its track record was characterised by a strategy of constant negotiation that precluded a position of greater confrontation with the City Council, and this had produced several internal conflicts over the years. Thus, we argue that this initiative could be an example of negotiated climate resilience that managed to live during almost three decades. However, the lack of an adequate leadership handover at the internal level caused the model to deplete. The association's board, avoiding any internal debate, negotiated with the administration the end of the public-community co-management process (Interviews 3 and 4), which was formalised in 2016, whereby the City Council took over the management of the garden. The municipal government took advantage of this decision to privatise the management of almost all the city's urban gardens.

Some of the questions arising from the long-term co-management process of the Miraflores Park point to the complexity involved in the public-community management of this activist project. For instance, key questions concern updating the decision-making organs and processes, the precariousness and institutional dependence due to the project's economic instability and the roles adopted by the people who carry out each of the functions (Pozo 2019). These questions can also be analysed from the gender and age perspectives.[3]

Except for a few cases, the different strategies adopted by each community initiative and, above all, the lack of adequate municipal support since 2012 led to a widespread shift from a public-community management model to a public-private model (see Table 9.1). Thus, a standardised and bureaucratised model affected the specific identities that had developed each of these processes. Professionals with no connection to the neighbourhood communities replaced the facilitating teams that had been socially and politically involved in the community processes. Consequently, the garden collective dissolved, and the community management of the space ceased to exist, as did the capacity for collective creation and transformation (Interview 1) undermining their potential for building resilience.

Nevertheless, other experiences that opted for different community management models remained (e.g. San Jerónimo and Rey Moro social gardens) and acquired greater independence from municipal power, enabling them to carry on their community management processes up to the present day (Interview 2).

As the interviewed actors themselves stated (Interviews 2 and 3), there are substantive differences between community management, public-community co-management and public-private co-management concerning to the construction of neighbourhood links and sense of collective ownership that are required conditions to negotiated climate resilience models:

> If it's only each of them looking at their own little piece of land and looking out for their own interests, well, in the end what you have is just the opposite, a mechanism to privatise and appropriate what is collective. (Interview 2)

> *La Casa de las Moreras* (in Miraflores Park) used to be a lively place. There were always people there, it had a heart and soul ... Today it's a loveless project. It's public but not communitarian. Everything that's communitarian is public, but not everything that's public is communitarian. No one there feels that it belongs to them. (Interview 3)

[3] In fact, a dialogue process is being promoted between the different visions of the conflict through a community memory recovery project (La Digitalizadora de la Memoria Colectiva 2021).

The model in which the administration provides the spaces and some of the resources, and the community movements self-organise, still generates some conflicts among different visions of what each concrete space should be. However, as the actors also recognise, these conflicts can enrich the process, generate dialogue and form part of the learning.

9.5 Conclusions

For centuries, Seville has faced severe hydro-climatic risks affecting the city with a strategy of local hydrography radical transformation framed in a top-down decision-making model. Carried out with little or no citizen involvement, urban climate resilience—in fact, *urban climate defence*—has been achieved through a robust and expensive infrastructure system. In this context, which conditions the problems and their potential solutions, an alternative approach is analysed, based on the construction of resilience as a bottom-up and nature-based process. This form of urban production reveals the commitment and determination of civil society to build contested spaces that contribute to a more climate-resilient city and to generate community resilience in contexts of institutional neglect.

This case study shows that the large-scale infrastructure works, especially for protection against flooding, have prioritised safeguarding the interests of the city's urban elites and the areas where they have settled. These interventions have generated a series of direct and indirect resilience trade-offs, as illustrated by the investments based on improving the living conditions in the new areas of the city where these elites reside, and the prioritisation of speculative interests that triggered urban expansion into areas of the city that were prone to floods.

In contrast, Seville's social geography shows that large numbers of working-class citizens are concentrated in neighbourhoods in the outskirts of the city, where community experiences as the ones presented here have emerged. Historically, factors of social vulnerability (poverty, unemployment, minorities) have overlapped with environmental injustices in these neighbourhoods, especially regarding poor-quality housing, infrastructure and public space, which have turned them into highly sensitive neighbourhoods to hydro-climatic risks, especially concerning heatwaves. This, together with a process of institutional neglect by the City Council regarding the provision of public infrastructure and services, and specifically green areas, has been the breeding ground of experiences of spatial contestation and radical resilience.

The discourse underlying these community demands, as explicitly expressed by one of the *citizen park committees*, the CPPEM (1992), justifies the need for green spaces as mechanisms to mitigate these socio-ecological vulnerability factors. Green spaces, as central components of green infrastructure, support urban resilience to hydro-climatic risks (Juvillà 2019), improve the living conditions of the areas in which they are located, become an instrument of empowerment and response to social problems and can be a form of co-production of the city.

Communities promote a way of living and creating the city that Lefebvre understood as spatial practices (1974), which contribute to strengthening the social fabric of the areas where they are produced and to articulating processes of reinforcing community resilience. These processes of self-organisation of civil society gave rise to transformations in urban space (*urbs*), as a result of social praxis and an urban experience constructed by its inhabitants (*civitas*) who, through contestation, intervention and negotiation with the structures of government (*polis*), configured what we could call contested and negotiated processes of urban and community resilience: experiences that again challenge the underlying assumption in many relevant policy documents that *it is the city government that is in the lead position to deliver urban resilience at the city scale* (Ziervogel et al. 2017, p.3).

Even so, the community management and public-community co-management processes are not conflict-free and therefore threatened by the continuity of the experiences and their projection over time. Collective creation is fragile and costly; it requires great commitment, which may generate burnout situations that, as we have seen, can be occasionally used by the institution to subvert those processes. In every case, there is a visible need for a team of people to facilitate the processes, promote the values and principles and drive the actions needed to construct a collective reality that generates community climate resilience.

Concerning infrastructure, over the past two centuries, the city has established a dominance and occupation relationship with the complex network of rivers and streams on which it sits. These watercourses have been profoundly altered and have lost their heritage and identity values (Zoido and Fernández 1996; García García 2004). In contrast, the strategy to integrate these courses (or what remains of them) into the city's major parks through a Green Belt is the social demand for a city model in which the *concept of integration replaces the concept of dominance* (García García 2004, p.22).

Acknowledgements This work received funding from the Spanish Ministry of Science and Innovation through the RESCITIES project: Plan Nacional (PGC2018-100996-A-I00 (MCIU/AEI/FEDER, UE)) and from the Andalusia Government via Consolidation Research Groups Funds (2019/HUM-396 GIEST). We also thank all the agents interviewed and persons who bring us materials for their willingness.

References

Abbar S, Zanouda T, Borge-Holthoefer J (2016) Robustness and resilience of cities around the world

Ameneiro A (2021) Anillo Verde para Sevilla: 20 entidades piden someter el proyecto a participación pública. Diario de Sevilla. https://www.diariodesevilla.es/sevilla/Anillo-Verde-Sevilla-entidades-participacion-proyecto-Plan-emergencia-climatica-sequia_0_1552046532.html Accessed 26 Jul 2021

Ameneiro A (2012) La gestión de los huertos del Tamarguillo pasará al distrito. Diario de Sevilla. https://www.diariodesevilla.es/sevilla/gestion-huertos-Tamarguillo-pasara-distrito_0_575642859.html Accessed 26 Jul 2021

Anguelovski I, Shi L, Chu E et al (2016) Equity impacts of urban land use planning for climate adaptation. J Plan Educ Res 36:333–348. https://doi.org/10.1177/0739456X16645166
Association Climatologica de Media-Garonna y del Sud du Sud-Ouest (ACMG) (n.d.) Adaptation au changement climatique et optimisation du confort thermique par l'utilisation de recours renouvelables
Bahadur A, Tanner T (2014) Transformational resilience thinking: putting people, power and politics at the heart of urban climate resilience. Environ Urban 26. https://doi.org/10.1177/0956247814522154
Berkes F, Ross H (2013) Community Resilience: Toward an Integrated Approach. Soc Nat Res Int J 26:1, 5–20. https://doi.org/10.1080/08941920.2012.736605
Brand FS, Jax K (2007) Focusing the meaning (s) of resilience: resilience as a descriptive concept and a boundary object. Ecol Soc 12:23
Campbell LK, Svendsen ES, Sonti NF, Johnson ML (2016) A social assessment of urban parkland: Analyzing park use and meaning to inform management and resilience planning. Environ Sci Pol 62:34–44. https://doi.org/10.1016/j.envsci.2016.01.014
Capel H (2003) Los problemas de las ciudades. Urbs, civitas y polis. Colección Mediterráneo Económico, ISSN 1698-3726, 3: 2003. (9–22)
Carmona J (2004) Del dicho al hecho… ¿hemos andado ese trecho? Veinte años de lucha ciudadana por el Parque Miraflores. In: Encina J et al. Democracias participativas e intervención social comunitaria desde Andalucía. Atrapasueños, Sevilla
Chandler D (2014) Beyond neoliberalism: resilience, the new art of governing complexity. Resilience 2:47–63. https://doi.org/10.1080/21693293.2013.878544
Chelleri L, Baravikova A (2020) Understandings of urban resilience meanings and principles across Europe. https://doi.org/10.1016/j.cities.2020.102985
Chelleri L, Schuetze T, Salvati L (2015) Integrating resilience with urban sustainability in neglected neighborhoods: challenges and opportunities of transitioning to decentralized water management in Mexico City. Habitat Int 48. https://doi.org/10.1016/j.habitatint.2015.03.016
Comité Pro Parque Educativo Miraflores (CPPEM) (1992) Programa de intervención Socio-educativa en el Parque de Miraflores. Comité Pro Parque Educativo Miraflores, Sevilla
Confederación Hidrográfica del Guadalquivir (CHG) (2021) Plan Hidrológico de la Demarcación del Guadalquivir 2022-2027, Esquemas de Temas Importantes, Anexo 6, Sevilla
Constenla E (1999) El Ayuntamiento de Sevilla regenerará 104 has de suelos degradados. Diario El País. https://elpais.com/diario/1999/02/06/andalucia/918256935_850215.html. Accessed 26 Jul 2021
Davies JS (2013) Rethinking urban power and the local state: hegemony, domination and resistance in neoliberal cities. Urban Stud 51:0042098013505158. https://doi.org/10.1177/0042098013505158
Davoudi S, Shaw K, Haider LJ et al (2012) Resilience: a bridging concept or a dead end? "Reframing" resilience: challenges for planning theory and practice. Interacting traps: resilience assessment of a pasture management system in Northern Afghanistan. Urban resilience: what does it mean in plan. Plan Theory Pract 13:299–333. https://doi.org/10.1080/14649357.2012.677124
De Carli B (2016) Micro-resilience and justice: co-producing narratives of change. Build Res Inf 44:775–788. https://doi.org/10.1080/09613218.2016.1213523
Del Moral IL (1991) La obra hidráulica en la cuenca baja del Guadalquivir (siglos XVIII-XX): gestión del agua y organización del territorio. Universidad de Sevilla, Sevilla
Delgado M (2004) De la ciudad concebida a la ciudad practicada. Arch Cuad crítica la Cult
Delgado Ruiz M (2017) El urbanismo contra lo urbano. La ciudad y la vida urbana en Henri Lefebvre. Revistarquis 7:65–71. https://doi.org/10.15517/ra.v7i1.31895
Díaz del Olmo F, Almoguera P (coords.) (2012) Sevilla, la ciudad y la riada del Tamarguillo (1961): inundación y renovación urbana en Sevilla. Universidad de Sevilla, Secretariado de Publicaciones, Sevilla

Díaz Parra I (2010) Sevilla, cuestión de clase. Atrapasueños, Sevilla

Dimuro G, Soler MM, de Manuel E (2013) La agricultura urbana en Sevilla: entre el derecho a la ciudad y la agroecología. Hábitat y Sociedad 6:41–60. https://dx.doi.org/10.12795/HabitatySociedad.2013.i6.01

Dornelles AZ, Boyd E, Nunes RJ et al (2020) Towards a bridging concept for undesirable resilience in social-ecological systems. Glob Sustain 3. https://doi.org/10.1017/sus.2020.15

EMASESA (2020) Informe del 1° taller de participación. Plan de Emergencia Climática. https://participacion.emasesa.com. Accessed 26 Jul 2021

Fazey I, Carmen E, Chapin FS et al (2018) Community resilience for a 1.5 °C world. Curr Opin Environ Sustain 31:30–40

García García A (2004) Potencialidades de la rehabilitación de cursos fluviales en el marco de los nuevos modelos urbanos. El ejemplo del arroyo Tagarete en Sevilla. Andalucía Geográfica 10:18–25

García Jerez A (2011) Micro-conflictos espaciales y habitus político de los grupos contra-hegemónicos. Nomadas Rev Crit ciencias Soc y jurídicas 31(3):1–17

González Dorado A (1975) Sevilla: centralidad regional y organización interna de su espacio urbano. Servicio de estudios del Banco Urquijo, Sevilla

Gramsci A (1999) Selections from the prison notebooks of Antonio Gramsci. Elec B. https://doi.org/10.2307/479844

Guerrero JJ, Cáceres F, Giménez F, Moreira, JM (2016) Servicios de regulación climática aportados por la vegetación urbana a la Ciudad de Córdoba. 1ª Parte: Fundamentos y Metodología. REDIAM. Consejería de Medio Ambiente y Ordenación del Territorio. Junta de Andalucía

Guichot y Parody J (1877) Memoria de las inundaciones de Sevilla en los meses de diciembre del año 1876 y enero de 1877. Gironés y Orduña, Sevilla

Gutierrez L, García G, García I (2017) 'Soluciones Naturales' para la adaptación al cambio climático en el ámbito local de la Comunidad Autónoma del País Vasco. Ihobe, Sociedad Pública de Gestión Ambiental. Gobierno Vasco

Harris LM, Chu EK, Ziervogel G (2017) Negotiated resilience. Resilience 3293:1–19. https://doi.org/10.1080/21693293.2017.1353196

Huang X (2019) Understanding Bourdieu - cultural capital and habitus. Rev Eur Stud 11. https://doi.org/10.5539/res.v11n3p45

Jiménez, N (2014) Sevilla duplica su verde en una década. El Correo de Andalucía. https://elcorreoweb.es/historico/sevilla-duplica-su-verde-en-una-decada-KAEC717814 Accessed 26 Jul 2021

Jon I, Purcell M (2018) Radical resilience: autonomous self-management in post-disaster recovery planning and practice. Plan Theory Pract 19:235–251. https://doi.org/10.1080/14649357.2018.1458965

Junta de Andalucía (2019) Plan andaluz para la prevención de los efectos de las temperaturas excesivas sobre la salud 2019. Servicio Andaluz de Salud, Sevilla

Juvillà Ballester E (Coord.) (2019) Renaturalización de la ciudad. Área de Territorio y Sostenibilidad de la Diputación de Barcelona. Colección Estudios. Diputación de Barcelona, Barcelona

Kaika M (2017) 'Don't call me resilient again!': the New Urban Agenda as immunology … or … what happens when communities refuse to be vaccinated with 'smart cities' and indicators. Environ Urban 29:89–102. https://doi.org/10.1177/0956247816684763

Klinenberg E (2018) Palaces for the people: how to build a more equal and united society. Waterstone, London

Kleinschroth F, Kowarik I (2020) COVID-19 crisis demonstrates the urgent need for urban greenspaces. Front Ecol Environ 2020. https://doi.org/10.1002/fee.2230

Koliou M, van de Lindt JW, McAllister TP et al (2020) State of the research in community resilience: progress and challenges. Sustain Resilient Infrastruct 5:131–151. https://doi.org/10.1080/23789689.2017.1418547

La Digitalizadora de la Memoria Colectiva (2021). https://ladigitalizadora.org Accessed 26 Jul 2021

Lara García A (2018) Agua y espacio habitado. Propuestas para la construcción de ciudades sensibles al agua. Ed. Universidad de Sevilla, Sevilla

Lara García A, Satorras M, Berraquero-Díaz L et al (2020) Conceptualización e instrumentalización de la vulnerabilidad en los planes de gestión frente a riesgos hidroclimáticos. Análisis comparado de los casos de Barcelona y Sevilla. In: XI Congreso Ibérico de Gestión y Planificación del Agua, Madrid, España

Lefebvre H (1974) The production of space, 1991st edn. Basil Blackwell, Oxford

Low SM (2009) Towards an anthropological theory of space and place. Semiotica 2009:21–37. https://doi.org/10.1515/semi.2009.041

Lyons AP, Bourdieu P, Nice R (1980) Outline of a theory of practice. ASA Rev Books 6. https://doi.org/10.2307/532672

MacKinnon D, Derickson KD (2012) From resilience to resourcefulness: a critique of resilience policy and activism. Prog Hum Geogr 37:253–270. https://doi.org/10.1177/0309132512454775

Martín A (1996) Sevilla (1872–1994), ciudad y territorio. De lo local a lo metropolitano. Fundación Colegio Oficial de Aparejadores y Arquitectos Técnicos de Sevilla. Sevilla.

Meerow S, Newell J (2016) Urban resilience for whom, what, when, where, and why? Urban Geogr 3638. https://doi.org/10.1080/02723638.2016.1206395

Meerow S, Newell JP, Stults M (2016) Defining urban resilience: a review. Landsc Urban Plan 147:38–49. https://doi.org/10.1016/j.landurbplan.2015.11.011

Meerow S, Pajouhesh P, Miller TR (2019) Social equity in urban resilience planning. Local Environ 24:793–808. https://doi.org/10.1080/13549839.2019.1645103

Meerow S, Stults M (2016) Comparing conceptualizations of urban climate resilience in theory and practice. Sustainability 8. https://doi.org/10.3390/su8070701

Moser S, Meerow S, Arnott J, Jack-Scott E (2019) The turbulent world of resilience: interpretations and themes for transdisciplinary dialogue. Clim Chang 153:21–40. https://doi.org/10.1007/s10584-018-2358-0

Mulligan M, Steele W, Rickards L, Fünfgeld H (2016) Keywords in planning: what do we mean by 'community resilience'?. Int Plan Stud 21:4, 348–361. https://doi.org/10.1080/13563475.2016.1155974

Palomo F B (1878) Historia crítica de las riadas o grandes avenidas del Guadalquivir en Sevilla desde su reconquista hasta nuestros días. Ed. facsímil (2001) Ayuntamiento de Sevilla, Sevilla

Parejo J (2021) Sevilla ante el cambio climático. Diario de Sevilla. https://www.diariodesevilla.es/sevilla/Sevilla-cambio-climatico_0_932006948.html. Accessed 20 July 2021. Accessed 26 Jul 2021

Pearson LJ, Newton PW, Roberts P (2014) Resilient sustainable cities: a future

Pozo R (2019) Huerta las Moreras: belleza y complejidad de la gestión vecinal. El Topo Tabernario 31. https://eltopo.org/huerta-las-moreras-belleza-y-complejidad-de-la-gestion-vecinal/?fbclid=IwAR2Ltu725Oomty6Q7k17IT0JXXWLegVnIwAPZLkoApgzyqR8sSCgAOR6rMU Accessed 26 Jul 2021

Satorras M, Lara A, Ruiz-Mallén I (2020) Booklet of urban resilience community initiatives in Seville and Barcelona: civil society against the effects of climate change. RESCITIES project "The political ecology of urban resilience to hydro-climatic phenomena in Spain". UOC research group TURBA. https://turbain3.files.wordpress.com/2020/04/rescities-booklet-english-9.9.20.pdf Accessed 26 Jul 2021

Shokry G, Connolly JJ, Anguelovski I (2020) Understanding climate gentrification and shifting landscapes of protection and vulnerability in green resilient Philadelphia. Urban Clim 31:100539. https://doi.org/10.1016/j.uclim.2019.100539

Solís RJ (2022) Las inundaciones en la Sevilla contemporánea (1801–2015). La acción de los poderes públicos (1801–2015). Diputación de Sevilla, Sevilla.

Vale LJ (2014) The politics of resilient cities: whose resilience and whose city? Build Res Inf 42:37–41. https://doi.org/10.1080/09613218.2014.850602
Vanney JR (1970) L'hydrologie du bas Guadalquivir. CSIC, Madrid
World Health Organisation (2016) Health as the pulse of the new urban agenda. United Nations Conference on Housing and Sustainable Urban Development, Quito 2016
Ziervogel G, Pelling M, Cartwright A et al (2017) Inserting rights and justice into urban resilience: a focus on everyday risk. Environ Urban 29:123–138. https://doi.org/10.1177/0956247816686905
Zoido Naranjo F, Fernández Salinas V (1996) Las relaciones ciudad-río en Andalucía. Estudio de su evolución reciente a partir del planteamiento urbanístico y territorial. In: II Jornadas de Geografía Urbana 1995. Alicante, Universidad de Alicante, pp 337–363

Part IV
Final Remarks

Chapter 10
Afterword: Transformation Pathways Within Urban Climate Resilience

Isabel Ruiz-Mallén

Abstract This last section of the book summarises and invites reflection on the limitations and potentials of both institutional and grassroots initiatives included in this book to build transformational processes when planning and implementing urban resilience (top-down and bottom-up) strategies towards climate change. After briefly reviewing the concepts related to transformation and learning commonly used in the urban (climate) resilience literature, with a focus on environmental justice and social equity, I highlight the main ideas and perspectives for critically approaching urban resilience and transformation that are developed through the case studies presented and discussed in previous chapters. In doing this, I try to encapsulate and reflect on the main challenges and opportunities for more transformative ways of building urban climate resilience from both government-led interventions and bottom-up efforts or through co-production schemes, considering factors and conditions shaping power dynamics and exacerbating social vulnerabilities in the different cities analysed throughout the book. I also rely on the compiled knowledge and experience from the presented case studies to envision and reflect on the ways forward to foster urban climate resilience from a transformational perspective, without overlooking inherent power issues.

Keywords Co-production · Equity · Grassroots · Learning · Transformative governance

I. Ruiz-Mallén (✉)
Estudis de Psicologia i Ciències de l'Educació and Internet Interdisciplinary Institute (IN3), Universitat Oberta de Catalunya (UOC), Barcelona, Spain
e-mail: iruiz_mallen@uoc.edu

I. Ruiz-Mallén et al. (eds.), *Urban Resilience to the Climate Emergency*, The Urban Book Series, https://doi.org/10.1007/978-3-031-07301-4_10

10.1 Transformation, Learning and Urban Climate Resilience

The New Urban Agenda advocates a set of transformative commitments in urban governance for enhancing climate resilience that imply *promoting international, national, subnational and local climate action, including climate change adaptation and mitigation, and supporting the efforts of cities and human settlements, their inhabitants and all local stakeholders as important implementers* (UN 2017, p. 22). Thus, this Agenda represents an important political step in terms of achieving an international consensus on the unsustainable growth and development of cities worldwide and its negative consequences (e.g. environmental degradation, growing inequalities, etc.) within and beyond their boundaries and the need for transformative changes in current policymaking.

In parallel, in the last few years, academic and practitioners' voices arguing that urban resilience approaches dealing with climate change must be linked to the concept of transformation have been increasing (Bahadur and Tanner 2014; Fazey et al. 2018; Elmqvist et al. 2019; see also Chap. 1 for a critical review of urban resilience from the lens of political ecology). From an environmental justice perspective, the reasons behind this understanding mostly refer to the necessity of engaging with power issues around resilience decision-making and dealing with trade-offs shaping social vulnerability (Ziervogel et al. 2017; see Chap. 2 for the main literature related to urban transformative adaptation and environmental justice). Thus, the concept of transformation in the urban climate resilience literature can be posed in terms of *deliberative transformation* (see also Chap. 4) that implies an intention and the recognition that key changes in traditional urban policymaking approaches are necessary to enable desirable futures by considering ethics and sustainability as priorities and without forgetting about power-related issues involved in the definition of what is desirable (O'Brien 2012; Davoudi et al. 2013).

As argued by Hölscher et al. (2019, p. 187), *urban transformations are complex, long-term, uncertain and contested processes of radical change in urban systems* so new forms of developing urban governance consisting of *new collaborations across domains, long-term planning horizons beyond electoral cycles, and learning-based approaches that allow for innovation and synergies to emerge and flourish* are required for transformation. These authors put the focus on governance capacities to highlight the need of enhancing institutions' learning capacities to anticipate and recover from climate perturbances while looking for desirable alternatives, for which approaches that integrate different types of knowledge are needed (Hölscher et al. 2019). However, as noted by environmental justice scholars, even when the root causes of vulnerability have been identified and participatory and collaborative governance and learning approaches are adopted by city governments, power imbalances can still influence urban resilience decision-making because citizens' multiple perceptions, needs and values and their historical barriers to access to institutions and infrastructures are not fully understood, nor integrated (Ziervogel et al. 2017; Anguelovski et al. 2020; Calderón-Argelich et al. 2021).

By relying on social-ecological resilience (Folke et al. 2010), Davoudi et al. (2013) put the social learning capacity of the system as a previous step for being transformative, for which collective memory and individual testimonies of past experiences related to climatic events are needed. O'Brien (2012) refers to the transformative learning processes identified in the literature on education. Transformative learning emerges from individual and collective critical reflection, revision and questioning of deeply held assumptions and predefined mental models leading to a conscious change in behaviours and social norms (Mezirow 1997) in order to recognise and integrate multiple perspectives and types of knowledge in climate change adaptation (Fazey et al. 2018). Also from the social-ecological systems literature, Fedele et al. (2019, p. 118) understand transformative adaptation to climate change as a strategy different from accommodating benefits to continue to provide benefits but as one that relies on *fundamental systemic changes that create new states and interactions within social-ecological systems*. These authors conceptualise transformative adaptation through six main characteristics of these strategies (i.e. restructuring, path-shifting, innovative, multiscale, system-wide and persistent) and related opportunities to deliver on them that rely on fostering multi-stakeholders and cross-sectoral collaborations, promoting multi-loop learning approaches, identifying power dynamics, questioning current assumptions on dominant values and practices, monitoring progress in the long term, engaging with multiple levels of governance and securing political support to ensure persistence in the long term, among others. Similarly, the discussion of the urban transformative capacities' framework by Wolfram et al. (2019) provides helpful guidance towards promoting transformation and learning in urban climate resilience research and practice. For instance, these authors reflect on the need to approach *agency from the focus on inclusion and empowerment as a bottom-line condition to the role of urban planning, local academia, and intermediaries in (re-)shaping inclusive and multi-form governance arrangements*, as well as further exploring those *common factors that shape transformative leadership* and the *social learning and political dynamics* behind innovations and experiments in urban climate governance.

Moreover, some authors claim for providing urban resilience thinking with normative guidance (Bahadur and Tanner 2014; Fazey et al. 2018). They argue that the often-absent definition of subjects when building resilience through city climate change plans and policies deters from identifying the winners and losers of the resultant actions, making it difficult to address potential trade-offs that negatively impact those already vulnerable communities. Further, these authors strongly suggest looking at how grassroots mobilisation towards urban climate resilience is contesting unequal top-down policies and putting pressure on urban government authorities. For instance, this is the case of the new climate movements towards climate emergency (see Chap. 8 for the case of the Barcelona climate emergency declaration). Important lessons for transformation can be learned from these community or citizen-based initiatives and projects in terms of creating partnerships, including marginalised voices and recognizing power asymmetries (O'Brien 2012; Cloutier et al. 2018). This way of looking at resilience from the grassroots should not be misinterpreted by using neoliberal approaches to reduce government responsibilities in terms of protecting the welfare of citizens and transferring these responsibilities to the community. On the contrary, it is a claim to foster citizens' collaborative engagement in resilience decision-making at the city level to move away from business as usual and make transformation an overarching goal.

This brief review on how transformation is understood and approached by recent literature dealing with urban governance and climate change resilience, particularly in relation with learning processes, provides a relevant basis for the next section to reflect on the approaches, challenges and opportunities for more transformative ways of building urban climate resilience (both top-down and bottom-up strategies) that are discussed in the case studies throughout this book. See an extended review of transformative adaptation and climate justice literature in Chap. 2.

10.2 Insights from the Book's Case Studies: Approaches, Challenges and Opportunities for Transformation

As presented above, the seven case studies included in this volume are structured into two parts (i.e. Parts II and III), as this section is divided as well.

10.2.1 Top-Down Climate Resilience and Transformation

Part II is focused on the analysis of top-down urban plans, policies and interventions related to climate resilience in different cities. It includes two chapters that provide historical and recent evidence from the city of Barcelona (Chaps. 3 and 4 led by Jordi Honey-Rosés and Isabelle Anguelovski, respectively) and a third one framed at continental level that covers the Latin American and the Caribbean (LAC) region (Chap. 5 led by Pere Ariza-Montobbio).

The three chapters conceptually deal with the concept of transformation, in some cases by using specific related terminology (e.g. transformational urban adaptation), although they also show important and interesting differences rooted in the understanding of the concept. Moreover, these chapters provide synergistic and interesting lessons in terms of the barriers and potentials for going beyond mainstream approaches in urban resilience policymaking. Findings from these chapters also emphasise the relevance of considering the geographical and historical context (and its different characteristics and conditions) when planning for and implementing climate resilience to learn from and address locally framed aspects related to social equity and environmental justice issues.

Jordi Honey-Rosés, in Chap. 3, *Urban Resilience in Perspective: Tracing the Origins and Evolution of Urban Green Spaces in Barcelona*, situates his work within an historical perspective to analyse how past urban planning decisions have led to the present configuration of the city green network and their implications in terms of green gentrification and social exclusion. In doing this, he reflects on how resilience planners can learn from past transformations to effectively deal with the current climate emergency while embracing a social justice perspective. His analysis perfectly dovetails with the examination of the ups and downs of the 2016 Poblenou Superblock project in the city as a recent political effort from the Barcelona municipality for promoting transformational land use planning and climate adaptation, elaborated by Isabelle Anguelovski, Christos Zografos, Kai Klause and James J.T. Connolly in Chap. 4, *Urban Transformational Adaptation: Contestation and Struggles for Authority in the Pilot Barcelona Superblock of Poblenou.* Prof. Anguelovski and colleagues use the concept of *transformational urban adaptation* applied to the Poblenou Superblock project understood as a climate adaptation intervention of transformational nature led by the city government. These transformational adaptation interventions are conceived as large-scale or intensity actions that reorder and transform cities by confronting *the underlying failures of development* (e.g. greenhouse gas emissions) and *by engaging with the politics of managing risk* to address climate change vulnerabilities and the uneven distribution of risk across the city (Kates et al. 2012; Ribot 2011; Connolly 2018). Dr. Honey does not use such terminology but argues that some historical interventions in the city were transformational too in this regard. For instance, he explains how the nineteenth-century Cerdà Plan was of transformational nature because Cerdà proposed *an equitable distribution of open spaces* across the city while *the mainstream approach of his time* consisted of *several large open spaces, parks, forests or gardens within the city fabric* (see Chap. 3).

In Chap. 5, *Urban Resilience in Latin America: Questions, Themes and Debates*, Pere Ariza-Montobbio, Andrea Carrión and Gian Carlo Delgado-Ramos bring the Latin American perspective on the connections between urban resilience policymaking and transformation in LAC region. By relying on the testimonies of key actors from the fields of urban resilience and climate change adaptation, these authors emphasise the need for structural transformation in urban resilience policy and practice. Their findings show how, *when the focus is on reactive action to disasters rather than structural transformations*, resilience practice is perceived *as following obsolete and maladapted practices for rebuilding urban habitats*. There is *a predominant biophysical view* of urban resilience among urban authorities in LAC, they argue, *focused on the technological recovery of infrastructures or with a neoliberal view focused on individual adaptive capacities*; no view considers *collective, societal and political power structures* that are found as central to build *a more systemic and transformative approach* in LAC. In this sense, transformation within the context of urban (climate) resilience in LAC seems to be first and foremost linked to structural interventions that secure basic urban services to guaranteeing people's basic rights such as *food sovereignty, public health, water and sanitation.* Such a call focused on basic human rights makes a difference in the approach to transformative ways of building climate resilience with the two previous chapters placed in Barcelona. In line with Ziervogel et al. (2017), Dr. Ariza-Montobbio and colleagues also suggest that *resilience is a concept defined and constructed in the global North that do not explicitly incorporate a contextualized view that recognizes the differences of what urban resilience may mean in the global South*. Consequently, the approaches undertaken to transformation when building urban climate resilience in LAC may also lack of appropriate contextualization and can suffer from such implementation gap, which is itself quite a challenge in LAC.

Related to this, the political instability and volatility of the region is documented by Dr. Ariza-Montobbio and co-authors as another key challenge for planning and implementing urban resilience by urban government authorities in LAC. By contrast, in the case of Barcelona, Chaps. 3 and 4 show how the city government is actively engaged in building climate resilience while trying to implement transformational adaptation interventions. Barriers in this case are not so much related to the role of city authorities but with the idea of *struggles for authority – understood as the legitimate ability to exercise and control power – may open up or close down spaces for transformational adaptation* (see Chap. 4). Prof. Anguelovski and colleagues argue that the Poblenou Superblock struggles for authority are not so much related to holding different visions for urban (and environmental) policy and planning as they are to *competitive urbanism and related short-term political gains*. The authors conclude with a novel contribution to the climate resilience literature by showing that deep urban transformations towards climate adaptation can be obstructed *because of the message they convey as concerns who has the authority to decide for the common good.*

Common challenges for city governments when implementing interventions for building climate resilience in public spaces, identified both in Barcelona and LAC, have to do with citizens' dissatisfaction about the social consequences of these interventions (e.g. social justice) as well as to how and how much they were involved in planning decisions. In Barcelona, Dr. Honey shows how the transformation of the *Turó de la Rovira* from an abandoned hill to an *educational and memorable* public green has attracted a – probably unexpected – *flood of visitors who appropriated the space and displaced locals, eclipsing their interests*. Prof. Anguelovski and co-authors show how neighbours' *discontent over public participation aspects* of the Poblenou Superblock project and their *perceived imposition of authority* were significant barriers for transformational adaptation when planning and implementing interventions from the top. In LAC cities, where documented participatory processes in Chap. 5 seem to be less socially inclusive than in the case of Barcelona, the participation challenge is mainly concerned with the inclusion of poor neighbourhoods and communities living in risk-prone areas in urban resilience decision-making. Dr. Ariza-Montobbio and co-authors notice that *participation should be under co-management schemes, going beyond consulting or coordinating exercises where technical experts and politicians take the most critical decisions* and put forward the example of *cLIMA sin riesgo in Lima, Peru,* as a project to make progress *towards this desirable participatory governance.*

City governments can learn from these highlighted historical and current barriers for transformational adaptation in Part II in order to enhance resilience and address social inequities in future urban planning for climate change. For instance, what it has been summarised here from Chaps. 3, 4 and 5 shows the importance of embracing participatory and inclusive approaches, generating contextual perspectives, considering multi-level as well as multi-sectoral and multi-actor policies in urban environmental and climate policy and planning as well as paying attention to socio-political conflicts and issues related to struggles for authority. The adoption of suitable measures to improve urban authorities' current practice in this regard depends on their institutional learning capacities. In this sense, the three chapters advocate for a (self-)critical revision of the most recent past actions to deal with climatic and other disturbances that can provide urban authorities with the tools to redirect their practice when needed. Dr. Honey argues that the adoption of a historical approach will allow recognising cyclical patterns of vulnerability and marginalisation and avoiding to reproducing them. Recent research on the Barcelona climate policies has noted how participatory and collaborative approaches for climate change policymaking are not always intensified over time (Satorras 2021). A longer timeframe than the one related to the implementation of the intervention itself is necessary to understand *how power is organized and how struggles for authority can be resolved*, as argued by Prof. Anguelovski and co-authors. In the case of the LAC region, it is necessary to go back to the colonial period, Dr. Ariza-Montobbio and co-authors remark, to understand how the *extractive industries and low-value-added services, frequently informal, contribute to a regional and urban development characterized by inefficient linear urban metabolic patterns* with resultant *inequalities, poverty and violence*. Being aware of this and adopting a contextualised and *sensitive approach to trade-offs and inequalities* is a necessary step for meaningful urban resilience.

The following section delves into the challenges and opportunities of the bottom-up initiatives analysed in Part III for more transformative climate resilience while summarising lessons learnt about how urban government authorities can better integrate learnings from the grassroots in urban climate planning and environmental governance. This is in line to what Dr. Ariza-Montobbio and co-authors conclude when referring to the work of Solecki et al. (2021) that social movements' and neighbourhood organizations' action to confront climate disasters as well as local government participatory planning processes have been shown to be key opportunities for building urban climate resilience with a transformative potential.

10.2.2 Bottom-up Climate Resilience and Transformation

Part III explores collaboratively and self-produced bottom-up initiatives dealing with climate change and other socio-ecological challenges in different urban areas. The initiatives analysed have been generated at community level as well as co-produced between local communities, urban government authorities and other organisations, which have helped to set into motion social learning processes. The topics covered by the four chapters included in this section are highly diverse. From the co-design and implementation of nature-based climate adaptation projects in schools (Chap. 6 led by Francesc Baró), the participatory schemes in wildfire governance projects (Chap. 7 led by Iago Otero) and the co-production of climate emergency action (Chap. 8 led by Mar Satorras) to the community-based and collaborative management of urban green spaces (Chap. 9 led by Ángela Lara García). Thus, these chapters include case studies focused on different analytical contexts (i.e. schoolyards, wildland-urban interface, the city and city green spaces), with a diversity of targeted actors (e.g. children, wildfire managers, new climate movements, working-class neighbourhoods) from different geographical settings (i.e. Paris, Barcelona, Madrid, Seville). The terminology used when referring to transformation is also diverse (e.g. nature-based transformation, radical transformation, social transformation). But interestingly, what they have in common is that their approach to transformation and urban climate resilience emphasises the need for reframing urban environmental governance within and beyond the city in two main aspects: (1) promoting deep socio-cultural changes at institutional level towards accepting that nature and society are intrinsically linked, and (2) recognising local communities' agency in urban climate resilience, as well as fostering their empowerment over the place, through more inclusive governance approaches that support social and transformative learning processes.

In terms of the challenges for building resilience with transformative potential, and similarly to the case studies analysing top-down interventions in Part II, these four chapters in Part III highlight how these bottom-up processes, particularly when they are co-produced, also require political and financial support from urban government authorities to guarantee their sustainability in the long term. For instance, in Chap. 6, *Nature-Based Climate Solutions in European Schools: A Pioneering Co-designed Strategy Towards Urban Resilience*, Francesc Baró, David A. Camacho, Carmen Pérez del Pulgar, Isabel Ruiz-Mallén and Pablo García-Serrano provide evidence from three pilot projects in Paris, Barcelona and Madrid showing how school settings have been transformed into *green environments* by using *co-production approaches, that is, engaging factual and potential users of these spaces since the earliest stages of the design process.* Among the challenges found, Dr. Baró and co-authors point out that the up-scaling of the project has not been planned in Madrid due to *lack of political support from the current municipal government*, while it has been planned to be implemented in other schools in Paris and Barcelona because, in these two cities, such kind of projects are *backed by municipal climate adaptation policies (Barcelona Climate Plan 2030 and Paris Resilience Strategy).* In Chap. 9, *Contested Spaces for Negotiated Urban Resilience in Seville*, Ángela Lara García, Luis Berraquero-Díaz and Leandro del Moral Ituarte analyse the historical claim of green spaces by working-class neighbourhoods in Seville and their achievements in terms of urban resilience and transformation (e.g. social garden projects). The authors reflect on how these grassroots initiatives were challenged when there was a change in the city government and the conservative party taking power significantly reduced the funding for these projects or stopped funding them. In turn, in Chap. 7, *Social-Ecological Transformation to Coexist with Wildfire: Reflecting on 18 Years of Participatory Wildfire Governance*, Iago Otero reveals that resources constraints impeded the long-term maintenance of *alternative land management projects* created from the grassroots (e.g. producers of artisan goat cheese, organic sheep/goat meat and firewood) in the Montseny (Barcelona metropolitan region) that could have resulted into a landscape with *less flammable* vegetation.

Chapters about co-produced urban climate resilience initiatives describe challenges for fulfilling their transformative potential that are related to procedural equity aspects. In Chap. 8, *Formal and Disruptive Co-production of the Climate Emergency Response: The Case of Barcelona*, Mar Satorras analyses this co-production process from the perspective of the city government leading a *formal* process of co-production and the perspective of climate movements that promote a parallel, *disruptive co-production* process. Climate movements questioned, among other aspects, who is allowed (and not allowed) to participate and make decisions in the co-production process led by urban authorities. In particular, they *criticized the engagement of business lobbies that were considered to represent private interests or the fossil fuel economy* and proposed *to target participants representing the general interest such as social agents and independent scientific experts*. In a similar vein, Dr. Baró and co-authors (Chap. 6) point out that the analysed nature-based climate adaptation projects excluded some potential users from the co-design process and did not prioritise social equity criteria when selecting participant schools in the co-production process, *at least in Paris and Barcelona* where the *socioeconomic characteristics of each district* were not considered.

Beyond these challenges, the analysis of the grassroots and co-produced efforts in urban climate resilience from diverse fields and sectors that are included in Part III also offer opportunities for more transformative ways of building resilience that particularly address environmental justice and social equity aspects. For instance, as argued in Chap. 6 by Dr. Baró and colleagues, future nature-based climate adaptation projects in schools should consider *higher exposure to climate harms (e.g., urban heat island) and socioeconomic characteristics* as key criteria for school selection, while securing enough resources to guarantee *the opening of the transformed schoolyards to the broader community* so that those more vulnerable citizens can benefit from these interventions. In the case of participatory wildfire governance (Chap. 7), Dr. Otero also reflects on inclusivity aspects related to the integration of *actors potentially affected by wildfire* and *their values* in the co-production process while putting the emphasis in the opportunities that *social cohesion* at the community level and beyond (i.e. among *public agencies, actors and citizens* involved in co-production) can bring in terms of collective learning to move forward transformative resilience. The author advocates for conceiving and managing social cohesion, and community cohesion in particular, as a crucial aspect for building resilience to wildfires because his findings show the *key role that a locally rooted and empowered civil society can have in the development of official resilience policies when it is able to network with key agencies and actors over the long-term*.

It is also remarkable that, in Chaps. 8 and 9, Dr. Satorras and Dr. Lara-García and co-authors, respectively for the cases of Barcelona and Seville, notice the relevant role that grassroots mobilisation and contestation have in shaping city governments' urban climate resilience agenda towards more transformative processes and outcomes. For instance, Dr. Satorras highlights how *the disruptive co-production process seemed to be key to incorporate radical demands from new climate movements potentially challenging powerful interests and established lobbies*, such as the *consideration of the Barcelona Port and Airport* within the scope of the city's Climate Emergency Declaration. Opening up these spaces for social contestation within *formal* processes of co-production in urban environmental and climate planning is therefore necessary for developing more transformative pathways within urban climate resilience.

10.3 Food for Thought

In this last chapter, I have tried to summarise the main aspects and issues raised throughout the volume regarding the limitations and potentials of both government-led interventions and grassroots initiatives to promote learning-based transformational processes when planning and implementing urban resilience strategies towards climate change.

Findings from the case studies developed in the seven chapters reviewed here (in Parts II and III) show how the analysed top-down, co-produced and self-managed climate resilience initiatives can face common challenges, such as the lack of political support. At the same time, these initiatives can suffer from specific limitations for transformation that are mostly related to procedural equity aspects. In the case of top-down interventions, for instance, citizens' perceived imposition of authority during participatory processes can demotivate them from getting more involved. In co-production processes organised by city governments, the promotion of inclusivity issues within such processes can lead to the presence of powerful actors in detriment of those more vulnerable in terms of hearing their voice.

Even so, the documented and analysed climate resilience initiatives can enact learning-based transformational processes for urban environmental and climate governance at different scales (e.g. schoolyards, neighbourhoods, cities, peri-urban areas), targeting a variety of actors (e.g. local residents, school communities, policymakers, social movements), and at different historical moments. In this last regard, some identified opportunities at research level are related to the analysis of the historical evolution of urban green space planning. Approaching urban climate resilience from a historical perspective can reveal valuable insights to fully understand barriers and limitations linked to urban collaborative governance practice, identify maladaptation and detrimental strategies and envision more transformative pathways for climate resilience in terms of inclusion and equity (Satorras et al. 2020; Satorras 2021). In the policy realm, findings from the case studies suggest that both government-led and co-produced climate resilience initiatives can benefit from building partnerships with social actors and coalitions to deal with socio-political barriers that are always present and thus promote multi-loop learning approaches while increasing public acceptance of innovative and transformational interventions.

In this line, there is a clear emphasis on co-production, relying on social and transformative learning and collaboration across sectors, as a strand of transformation (Elmqvist et al. 2019; Hölscher et al. 2019; Ruiz-Mallén 2020). But the different ways in that co-production can be approached, as showed, for instance, in Chap. 8 where Dr. Satorras makes the difference between *invited spaces* or *insisted spaces*, have direct implications in terms of climate justice, social equity and social learning. Consequently, it is relevant that further research on urban climate resilience critically examine how such collaborative interaction between citizens, urban government authorities, private sector and other actors is constructed and developed in order to understand who is involved or excluded, has voice or is ignored and therefore can learn or not as well as for whom resilience has better and worse outcomes. Moreover, as highlighted by the literature on urban climate resilience (O'Brien 2012), as well as by the book's chapters as a common message from the case studies, the identified structural and procedural barriers to transformation linked to politics, power and interests need a transdisciplinary approach to urban climate resilience.

In sum, the case studies showcased in this volume do not offer blueprint solutions to overcome these diverse challenges but provide interesting lessons to make some progress towards more socially inclusive urban climate resilience approaches in different urban contexts. There are many other interesting and useful insights on urban climate resilience and learning-based transformation that have resulted from the case studies. I encourage readers interested in these topics to return to previous chapters because each of them contains novel and fascinating conceptual and methodological reflections in this regard that are informed by specific topics and settings. The volume in general, as well as this last chapter conceived as an effort of synthesis, seeks to be a point of departure for future work to be made by those readers particularly concerned with the book's main topic, that is, *unravelling the transformative potential of institutional and grassroots initiatives*.

Acknowledgements I thank the financial support from the Spanish Ministry of Science and Innovation through the RESCITIES project: Plan Nacional (PGC2018-100996-A-I00 (MCIU/AEI/FEDER, UE)) and the Spanish State Research Agency through the grant RYC-2015-17676 funded by MCIN/AEI/ 10.13039/501100011033 and by "ESF Investing in your future".

References

Anguelovski I, Brand AL, Connolly JJT, Corbera E, Kotsila P, Steil J, Argüelles Ramos L (2020) Expanding the boundaries of justice in urban greening scholarship: toward an emancipatory, antisubordination, intersectional, and relational approach. Ann Am Assoc Geogr 110(2):1–27

Bahadur A, Tanner T (2014) Transformational resilience thinking: putting people, power and politics at the heart of urban climate resilience. Environ Urban 26(1):200–214. https://doi.org/10.1177/0956247814522154

Calderón-Argelich A, Benetti S, Anguelovski I, Connolly JJT, Langemeyer J, Baró F (2021) Tracing and building up environmental justice considerations in the urban ecosystem service literature: a systematic review. Landsc Urban Plan 214:104130

Cloutier G, Papin M, Bizier C (2018) Do-it-yourself (DIY) adaptation: civic initiatives as drivers to address climate change at the urban scale. Cities 4:284–291

Connolly JT (2018) From systems thinking to systemic action: social vulnerability and the institutional challenge of urban resilience. City Community 17(1):8–11

Davoudi S, Brooks E, Mehmood A (2013) Evolutionary resilience and strategies for climate adaptation. Plan Pract Res 28(3):307–322

Elmqvist T, Andersson E, Frantzeskaki N, McPhearson T, Olsson P, Gaffney O, Folke C (2019) Sustainability and resilience for transformation in the urban century. Nat Sustain 2:267–273. https://doi.org/10.1038/s41893-019-0250-1

Fazey I, Schäpke N, Caniglia G, Patterson J, Hultman J, van Mierlo B, Säwe F, Wiek A, Wittmayer J, Aldunce P, Al Waer H, Battacharya N, Bradbury H, Carmen E, Colvin J, Cvitanovic C, D'Souza M, Gopel M, Goldstein B, Hämäläinen T, Harper G, Henfry T, Hodgson A, Howden MS, Kerr A, Klaes M, Lyon C, Midgley G, Moser S, Mukherjee N, Müller K, O'Brien K, O'Connell DA, Olsson P, Page G, Reed MS, Searle B, Silvestri G, Spaiser V, Strasser T, Tschakert P, Uribe-Calvo N, Waddell S, Rao-Williams J, Wise R, Wolstenholme R, Woods M, Wyborn C (2018) Ten essentials for action-oriented and second order energy transitions, transformations and climate change research. Energy Res Soc Sci 40:54–70

Fedele G, Donatti CI, Harvey CA, Hannah L, Hole DG (2019) Transformative adaptation to climate change for sustainable socialecological systems. Environ Sci Policy 101:116–125

Folke C, Carpenter SR, Walker B, Scheffer M, Chapin T, Rockström J (2010) Resilience thinking: integrating resilience, adaptability and transformability. Ecol Soc 15(4):20

Hölscher N, Frantzeskaki N, McPhearson T, Loorbach D (2019) Capacities for urban transformations governance and the case of New York City. Cities 94:186–199

Kates RW, Travis WR, Wilbanks TJ (2012) Transformational adaptation when incremental adaptations to climate change are insufficient. Proc Natl Acad Sci 109(19):7156–7161. https://doi.org/10.1073/pnas.1115521109

Mezirow J (1997) Transformative learning: theory to practice. New Dir Adult Contin Educ 1997(74):5–12

O'Brien K (2012) Global environmental change II: from adaptation to deliberate transformation. Prog Hum Geogr 36(5):667–676

Ribot J (2011) Vulnerability before adaptation: toward transformative climate action. Glob Environ Chang 21(4):1160–1162

Ruiz-Mallén I (2020) Co-production and resilient cities to climate change. In: Narez J, Bole D (eds) Participatory research and planning in practice. The urban book series. Springer, Cham, pp 1–11

Satorras M (2021) From commissioned reports to co-production processes: the evolution of urban climate policy-making through the case of Barcelona. In: Leal Filho W, Luetz JM, Ayal D (eds) Handbook of climate change management. Springer, Cham, pp 4379–4401

Satorras M, Ruiz-Mallén I, Monterde A, March H (2020) Co-production of urban climate planning: insights from the Barcelona Climate Plan. Cities 106:102887

Solecki W, Delgado G, Roberts D, Rosenzweig C, Walsh B (2021) Accelerating climate research and action in cities through advanced science-policy-practice partnerships. NPJ Urban Sustain 1:3. https://doi.org/10.1038/s42949-021-00015-z

UN (2017) New Urban Agenda. United Nations, Quito

Wolfram M, Borgström S, Farrelly M (2019) Urban transformative capacity: from concept to practice. Ambio 48:437–448

Ziervogel G, Pelling M, Cartwright A et al (2017) Inserting rights and justice into urban resilience: a focus on everyday risk. Environ Urban 29(1):123–138. https://doi.org/10.1177/0956247816686905

Index

I. Ruiz-Mallén et al. (eds.), *Urban Resilience to the Climate Emergency*, The Urban Book Series, https://doi.org/10.1007/978-3-031-07301-4

The manufacturer's authorised representative in the EU is Springer Nature Customer Service Centre GmbH, Europaplatz 3, 69115 Heidelberg, Germany. If you have any concerns regarding our products, please contact ProductSafety@springernature.com

Printed and bound by CPI Group (UK) Ltd, Croydon, CR0 4YY
30/09/2026
02225493-0004